BASIC
WIRING
TECHNIQUES

Created and designed by the
editorial staff of ORTHO Books

Project Editor
Ken Burke

Writer
T. Jeff Williams

Designers
Craig Bergquist
Christine Dunham

Illustrator
Ron Hildebrand

Photographer
Fred Lyon

Photographic Stylist
Sara Slavin

Ortho Books

Publisher
Robert L. Iacopi

Editorial Director
Min S. Yee

Managing Editors
Anne Coolman
Michael D. Smith

System Manager
Mark Zielinski

Senior Editor
Sally W. Smith

Editors
Jim Beley
Diane Snow
Deni Stein

System Assistant
William F. Yusavage

Production Manager
Laurie Sheldon

Photographers
Laurie A. Black
Michael D. McKinley

Photo Editors
Anne Dickson-Pederson
Pam Peirce

Production Editor
Alice E. Mace

Production Assistant
Darcie S. Furlan

National Sales Manager
Garry P. Wellman

Operations/Distribution
William T. Pletcher

Operations Assistant
Donna M. White

Administrative Assistant
Georgiann Wright

Address all inquiries to:
Ortho Books
Chevron Chemical Company
Consumer Products Division
575 Market Street
San Francisco, CA 94105

First Printing in March 1982.

3 4 5 6 7 8 9

84 85 86 87

ISBN 0-89721-000-X

Library of Congress Catalog Card
Number 81-86181

Chevron Chemical Company
575 Market Street, San Francisco, CA 94105

Acknowledgments

Consultants:
Robert Beckstrom
Owner/Builder Center
Berkeley, CA

Kent P. Stiner
Professional Engineer
Traverse City, MI

**Front Cover
Photographer:**
Fred Lyon

Typography:
Vera Allen Composition
Castro Valley, CA

Color Separation:
Colortech
Redwood City, CA

Copyediting:
Carol Westberg
San Francisco, CA

Photographs:
Front Cover—Boxes provide
a place to connect wires to
other wires or to switches or
outlets.

Back Cover—Learn about
choosing the right kinds of
wires, fuses, and breakers;
tackling small repair jobs; or
wiring whole rooms.

Title Page—An assortment of
switches, boxes, and
receptacles.

BASIC WIRING TECHNIQUES

THE WIRING IN YOUR HOME

Before you start a project, check your grasp of the basics—like safety, identifying circuits in your house, and how to read a meter. This chapter will help you select quality tools for any job.

Electricity flows almost unnoticed into most modern homes. When you walk into a dark room, you just reach out, flip on a wall switch, and the lights go on—usually, anyway. Only when they don't, or when you plug in a heater and the room goes dark, do you stop to realize that electricity is no longer flowing through the mysterious wires hidden behind the walls. At this point, you go in search of the flashlight and make your way to the fuse box. You eventually find that a fuse was blown or that a breaker tripped. You correct the situation, and the lights go back on. Problem solved.

But what made them go out in the first place? What blew that fuse? "Just an overload," you say. "I'll be careful next time." That's one solution. A better solution might be to add a whole new electrical circuit to the house so you can safely run that heater and the television at the same time. But such a task appears formidable. After all, how can you add a circuit when you're really not sure what one is? Putting in a new circuit is not easy, but not impossible either. Hundreds are installed across the country every day by people no smarter than you.

Common Concerns

The most common concern of the home electrician is with the dangers of electricity. It can kill you, right? Or burn down the house? True, but if you understand how electricity works, use common sense, and take reasonable precautions, you will be in no danger (see the safety checklist on the next page).

Many people who would like to do some basic wiring in their homes believe that they can't do anything unless they hire a licensed electrician. Is it legal to do your own work? Yes, if you have a permit from your local building inspector (see page 13). Whenever you install something new to your wiring rather than just replacing something old, you need a permit. When the job is finished, an inspector will check your work to see that you have done it correctly.

Insurance is another common concern. Doing your own electrical work should not affect your house insurance. Nothing in a typical policy says that you cannot do your own work. But if a fire in your home were traced directly to poor wiring—which you did—the company might be reluctant to renew your policy.

Perhaps the most pervasive concern among home electricians is the belief that electricity is just too complicated. After all, look at all those wires! How are you supposed to know where they go? If you put one in backward, will the house blow up? Such feelings are somewhat justified. There are a lot of wires. But one nice thing about electricity is that it follows a definite logic. The wires have to go in the right sequence. If they don't, your house won't blow up, it just won't light up.

A Step-by-Step Approach

This book is about basic wiring, and it explains everything one step at a time. Once you understand how to replace different types of fuses, switches, and outlets, you will be ready to move along to more complicated projects. You will be able to add a few more outlets in a room or hallway and even to install a whole new circuit. All these steps in basic wiring are accompanied by wiring diagrams that show you the correct path of electricity.

This book covers how to fix common problems, such as a broken lamp or a doorbell that doesn't ring. And for the more ambitious, it describes basic wiring techniques, such as how to install a new ceiling fixture, including details on cutting through any type of ceiling in order to complete the electrical hookup. Finally, in the section on outdoor wiring you can learn how to extend circuits so you can enjoy your yard at night.

Remember, wiring is logical: you start at the beginning and work your way to the end. This book is designed to guide you step by step along that path.

◀

Electricity can add beauty and convenience to every room in your home.

PRINCIPLES OF ELECTRICITY

Safety and Common Sense

Electricity can be dangerous. But if you use common sense and follow prescribed safety measures, you will come to no harm. The basic rule is: *always shut off electrical power to the area in which you are working.* Then make sure it is off by using a voltage tester before touching any wires. If there is no electricity in the wires, you cannot be hurt. Here's a checklist:

■ Remove the fuse or turn off the breaker that controls the circuit you are working on. Post a sign on the box to warn others not to touch it because you are working on a circuit.
■ Before touching any wires, make sure they are dead by checking with a voltage tester (see page 11).
■ Never stand on a damp floor when working with electricity. If the floor around your fuse or breaker box is even occasionally wet, keep some boards or a rubber mat there to stand on before touching the box.
■ Never touch any plumbing pipes or gas pipes when working with electricity. They are often used as the ground to the electrical system. Touching a hot wire and the pipe at the same time could cause current to flow through your body.
■ Avoid using aluminum ladders near overhead entrance wires.
■ After completing your electrical work, turn the power on and use a voltage tester to check your work (see page 11). The tester should light when a connection is made between the hot black wire and the grounded box; it should not light when a connection is made between the white wire and the grounded box.

In addition to the basic safety procedures for working with electricity, follow these household safety tips:
■ Never use a fuse with an amperage rating higher than that specified for the circuit.
■ If small children may be in the house, keep receptacles (outlets) covered or install some type of safety receptacle to keep a child from poking a metal object into a slot (see drawing).
■ Always unplug a lamp before attempting any repairs.
■ Always pull on the plug itself, not on the cord. Teach your children to do this also.
■ Install smoke detectors in the house and garage.
■ When using an adaptor for a three-prong plug, make sure the adaptor is properly grounded by connecting it to the wallplate screw in the receptacle.
■ Immediately repair a cracked plug or worn electrical cord.
■ Avoid running extension cords across doorways or other traffic corridors.
■ Never run an extension cord under a rug where it can become worn and expose bare wires to flammable material.
■ Never touch faucets or other grounded objects while holding an electric razor, hair dryer, or other appliance.
■ When using power tools outdoors, or on concrete floors in contact with earth, make sure the electrical outlet is GFCI-protected (see page 23). If your outlets do not have GFCI protection, you can purchase a unit to wire into a regular outlet.

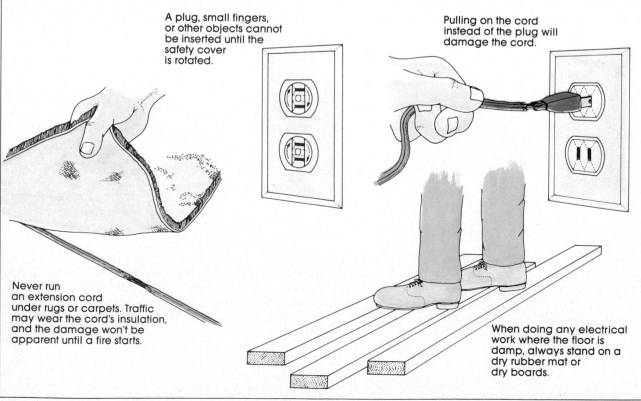

A plug, small fingers, or other objects cannot be inserted until the safety cover is rotated.

Pulling on the cord instead of the plug will damage the cord.

Never run an extension cord under rugs or carpets. Traffic may wear the cord's insulation, and the damage won't be apparent until a fire starts.

When doing any electrical work where the floor is damp, always stand on a dry rubber mat or dry boards.

The voltage and amperage or wattage requirements it will use are marked on every electrical device and appliance.

How Electricity Works

Many people are reluctant to start even a small home wiring project because they don't feel they understand how electricity works. Although the parallel is not exact, electricity is similar to water flowing through a hose. The flow of electricity must also be contained, for our purposes, in a wire. The moving electricity creates a current, called *amperes,* and it moves under pressure, known as *voltage.* When you multiply these two elements together, you get the number of *watts,* which indicates how much energy is being used. Grasping these three basic measurements of electrical power is essential to an understanding of electricity.

Amperes, commonly called amps, measure the amount of electrons flowing past a given point each second. As the flow of water is measured in gallons, electrical current is measured in amperes. About 6.28 billion billion electrons pass a given point each second to make 1 ampere, which is barely enough to ring a doorbell.

Volts tell you how much pressure is being used to push the electricity through the wires. Voltage, then, can be compared to water pressure. If you blow water through a straw, it will not spray very far, but the pressure in a fire hose will send water up several stories. The same principle applies in voltage. The 1½ volts in a flashlight battery will not hurt you, but 120 volts in an outlet can be fatal. Because voltage is a measurement of pressure, it is subject to resistance. The longer the wire, the more resistance, which weakens the voltage.

Voltage, as it arrives in your house, is subject to fluctuations from 114 to 126 volts. Although the National Electrical Code calculates electricity on the basis of 115 volts and 230 volts, many professional electricians refer to voltage as being either 120 or 240, which is a median of the fluctuations. This book also rates voltage at 120 or 240 throughout.

Watts designate the amount of power being used. Amperes alone do not tell you how much power there is, nor do volts alone. To find watts, multiply amps times volts. The number of watts per hour tells you how much energy an electric bulb or appliance is using.

Amps and volts can vary considerably and still provide the same number of watts. For example, if the headlight in your car draws 5 amperes from the 12-volt battery, it uses 60 watts (5 × 12 = 60). Likewise, if you plug a lamp with a standard bulb, drawing only ½ ampere, into your 120-volt house circuit, it also uses 60 watts (½ × 120 = 60).

A watt is a unit of power that is being used at any given moment. To calculate your energy usage, the power companies keep track of *watt hours.* Since one watt per hour is too minuscule to bother with, you are billed in *kilowatt hours,* which is abbreviated on your bill as *kwh.* The words *kilowatt* and *megawatt* are often bandied about by electricians. If you find them confusing, just remember that *kilo* stands for thousand, and *mega* stands for million. So a kilowatt is a thousand watts, and a megawatt is a million watts.

From Generator to Your House

Electricity originates in huge generators that are powered by water in hydroelectric dams or by coal, oil, or nuclear material. From the generator, electricity travels in high-voltage wires to distribution stations. From there it is directed into cities and towns, transformed into lower voltage, and finally lowered again to 120 volts before it enters a house. Electricity may be brought to your house in two wires or three wires.

If the house was built before 1940 and the wiring has not been modernized, you will see only two wires leading from the pole to the meter box. Each of the two wires carries 120 volts. One is a "hot" wire, which means it is carrying ungrounded electricity (see page 21). Hot wires are dangerous. The other is a "neutral" wire that completes the electrical circuit. Although technically neutral because it is grounded, it still carries current and must be handled carefully in case it is hot. Hot wires are normally black or red, and neutral wires are normally white (exceptions are explained at appropriate points in the book). However, never trust the wire color to be an accurate indicator of hot or neutral wires; trust only your voltage meter.

If you operate many modern appliances with high amperage demands, a two-wire system is probably inadequate for your house. In modern houses, three wires lead from the pole to the meter, supplying both 120 and 240 volts. Two of the three wires going to the house are hot, and the third one is neutral. The two incoming 120-volt wires are combined at one point in the service entrance or service panel to give you 240 volts for such major appliances as a range or dryer.

Service Heads

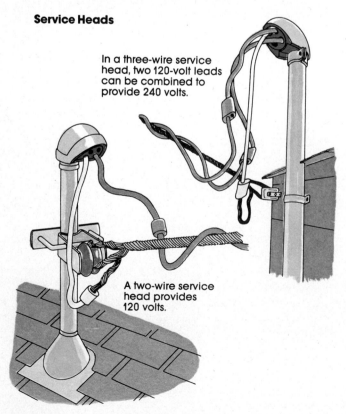

In a three-wire service head, two 120-volt leads can be combined to provide 240 volts.

A two-wire service head provides 120 volts.

Meter and Service Entrances

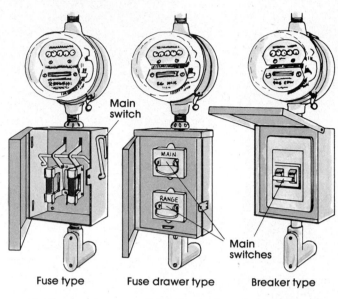

Main switch

Main switches

Fuse type   Fuse drawer type   Breaker type

In the service panel, the incoming electricity is distributed to the individual circuits in the house. Service panels are of two basic types: one with a circuit breaker and one with fuses (or fuses and cartridge fuses).

You can disconnect all power to the house by either throwing a switch, pulling a lever down, or pulling a main fuse block at the service entrance. If your house has a circuit-breaker panel, the main disconnect switch is normally located at the top of the panel just above the individual circuit breakers. If your house has a fuse box, two types of main power disconnects are possible. In one type, two cartridge fuses are in a box, which you simply pull out of the service entrance to cut the power. In the other type, you pull down a lever on the side of the box to disconnect the main power. Both fuses and circuit breakers allow you to disconnect the main power safely without being exposed to live wires.

Identifying Circuits

If all the circuits in your house are not already identified in the service panel, do it now. It's a first step in basic wiring. Start by taping a sheet of lined paper inside the cover. Consecutively number each fuse or breaker switch in the panel. This tells you how many circuits you have wired to the panel. Write those numbers in order on the paper.

Now turn on all the lights in your house, turn on any radios and televisions, and plug small lamps or night lights into the receptacles (outlets). Do not turn on major electric appliances such as stoves or dryers since they are on 240-volt circuits that you can check later.

At this point, you will need a helper. Unscrew one fuse or trip one breaker and have someone in the house tell you where the lights went off. Describe that area beside the matching number on the sheet of paper. In a two-story house, one circuit commonly goes to both floors, or at least two circuits serve each floor. That way, a blown fuse won't darken an entire floor. Make sure your assistant checks both floors every time you break a circuit.

Service Panels

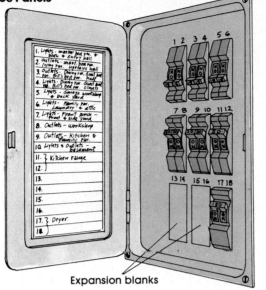

Expansion blanks

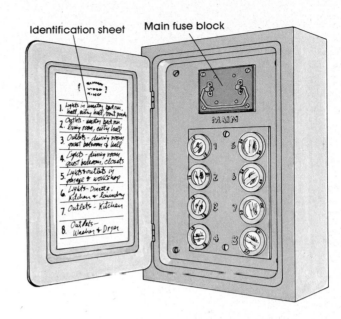

Identification sheet   Main fuse block

Continue in this manner until you have identified all the circuits in the house.

This map will help you quickly locate the trouble area when something blows a fuse or trips a breaker. And if you need to cut power to a certain section of the house for some repair work, you immediately know which fuse or breaker will do the job.

Determining a Circuit Overload

When a fuse blows or breaker switch trips, it usually indicates too much electrical demand on a particular circuit. To calculate the load a circuit will bear, add up all the watts used by lights and appliances that are commonly in use on that circuit at the same time. Watts are printed on bulbs and on a plate attached to any appliance or electric tool. Divide the total amount of watts by 120 volts to get the amperes. That figure should not exceed the total amperage allowed on that circuit. The amperage capacity is printed on the fuse or breaker switch. For example, a 15-amp capacity on a 120-volt circuit permits only 1,800 watts (15 × 120). If you exceed 1,800 watts, you overload the wires. *Never replace a fuse that is often being blown with one that has a higher amperage rating than indicated for the circuit.* If you use a 20-amp fuse on your 15-amp circuit, for instance, you could cause the wiring to overheat and start a fire.

Reading a Meter

Between the incoming power to your house and the service panel is a meter that keeps track of how much electricity you use. The meter records kilowatt hours on five dials. Some meters have dials similar to the mileage indicator on a speedometer. More commonly the dials resemble clocks with numbers running from 0 to 9.

Note that the hand on the far-left dial turns clockwise, the next one counterclockwise, the next one clockwise, and so on. Generally, the hand lies between two numbers. When reading a meter, always select the lower number, even if the hand is almost on the higher number.

If the hand is directly on a number, check the hand on the dial to the immediate right. If that hand has passed 0, then use the number the hand is pointing to. If the hand on the dial to the right has not passed 0, then use the next lower number. The dial to the right of any other dial must make one complete revolution before the one to the left moves up one digit.

In these examples, both meters read 50698. Note that the hand on the far-left dial points directly at the 5, but the hand on the fourth dial has just passed 0, indicating that the higher number should be used.

Take your own meter readings periodically to check your bill against the electric company's readings. Write down the meter readings the same time the meter reader checks your house. Do it for two consecutive months and then subtract the first from the second reading to find your consumption. If there are any notable discrepancies, discuss them with the electric company.

How to Read a Meter

KILOWATT HOURS

Meters register kilowatt hours and run continuously from when they are installed. If the needle on a dial is between two numbers, read the lower one. If the needle appears to be directly on a number, look at the next dial to the right. If that needle has not yet reached zero, the other needle has not yet reached the number. Both meters show the same reading, 50698.

TOOLS

You will be relieved to know that even for extensive wiring projects, you don't need many tools. And those you do need are not costly. You can do a great deal with nothing more than a screwdriver and a pair of pliers. But to make jobs easier and more pleasant, it's worth investing in many or all of the tools shown on this page. Many wiring jobs, such as putting in new ceiling lights or replacing switches, can be done with just a few tools. You probably have some of them in your household toolkit: several sizes of screwdrivers (standard and Phillips), hammer, pocket knife, and flashlight.

With the tools shown here, you can tackle almost any wiring job in your home. Some of them you will need only for specific jobs. You might need a hacksaw, for instance, only if you had to work with armored cable. But others will make many jobs easier and faster. For example, you can strip and cut wire with a knife and a pair of pliers, but a multipurpose tool will save you a lot of time and frustration. Since tools are an investment, always buy the best quality. Cheap tools are not worth the money; they will soon break and then you just have to buy others.

Tools

Slot screwdrivers

Phillips screwdrivers

Cold chisel

Wood chisel

Needlenose pliers

Lineman's pliers

Multipurpose tool

Aviation snips

Voltage tester

Continuity tester

Pocket knife

Cable stripper

Electrician's tape

Hammer

Electric drill

Spade bit

Bit extension

Flashlight

Fish tape

Hacksaw

Mini hacksaw

Compass saw

Using a Voltage Tester

A voltage tester is a miniature electrical circuit. Power comes in one wire, travels through the bulb, and goes out the other wire. Use it to make the following tests for safety and electrical continuity:

Testing wall receptacles. To test whether a wall receptacle is working or not, put one wire in each slot. The glowing bulb indicates power is there. If you plan to work on that receptacle, turn off the power to the circuit by tripping the proper breaker or removing the correct fuse. Check again with the voltage tester to make sure the power is off and then remove the receptacle (see page 44).

Finding a hot wire. Once a switch or receptacle has been removed from the box, you may forget which of the two or more black wires is carrying the current. To find out, bend the wires out so they are not touching each other or the box. Turn the power back on and put one probe on the metal box and the other on the exposed end of a black wire. The wire that causes the voltage tester bulb to glow is the hot wire. Turn the power off again and mark that wire with a piece of tape. If the box is plastic, touch one tester probe against the bare ground wire in the box and the other probe to a black wire.

Checking for grounding. The voltage tester is also used to check if a receptacle is grounded properly. To do this,

put one probe in the hot slot (the shorter slot on newer receptacles) and touch the other probe to the bare metal cover plate or the plate-holding screw. (If the screw is painted, scrape away enough of the paint until you can touch bare metal.) If the tester does not light up or shows a very weak light, the receptacle is not properly grounded. It is very important that the tester light just as strongly in the ground test as when you insert both probes in the slots.

Many receptacles have a third, U-shaped hole above or below the two slots. In these three-prong plugs, the U-shaped slot is the grounding slot. To test that this is working, insert one probe on the tester in the hot slot and the other probe in the grounding hole. *Important: the tester must light just as strongly in the ground test as it does when you insert both probes in the slots.* If it doesn't, the receptacle is not properly grounded.

Testing for power at a switch. To check whether power is on at a wall switch, first remove the wall plate. Put one probe on the metal box in the wall and the other probe on each of the terminals in the switch. If the tester lights on any terminal, there is power to the switch. If the box is plastic, touch one probe to the bare ground wire and the other probe to each terminal. If the tester lights, the switch is live. Turn off power to that point before doing any further work.

Using a Voltage Tester

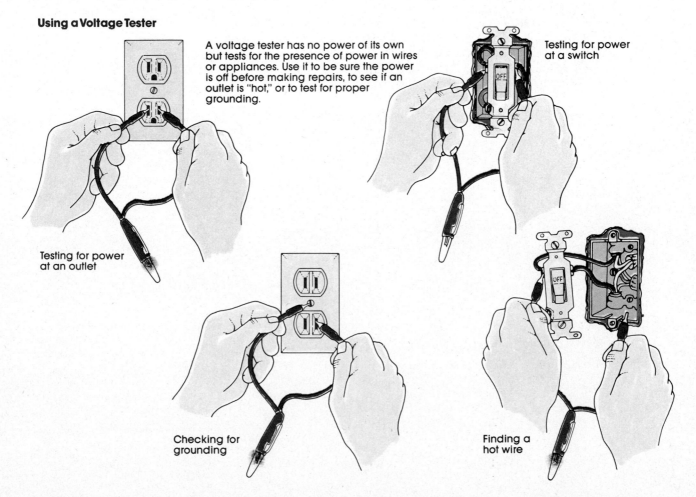

A voltage tester has no power of its own but tests for the presence of power in wires or appliances. Use it to be sure the power is off before making repairs, to see if an outlet is "hot," or to test for proper grounding.

Testing for power at an outlet

Testing for power at a switch

Checking for grounding

Finding a hot wire

Using a Continuity Tester

A continuity tester is used to check that a circuit is complete or continuous without having to turn on the house current. If the circuit is complete, the small battery in the tester will light up the bulb. If the circuit has been interrupted, the bulb will not light. Because it has its own power system, this device must never be used where electrical current is flowing. Use the continuity tester to check out any of the following problems:

Fuses. If you can't see that a fuse is blown, test it by putting the probe against the metal contact point at the base of the fuse and the alligator clip against the metal spring contact on a Type S fuse or on the metal threads of an Edison-base fuse (see page 28). If the fuse is good, the bulb will light.

Cartridges. Since you can't see when a cartridge fuse is blown, you must use a continuity tester. Check it by putting the probe on one end and the alligator clip on the other end. If the cartridge is good, the bulb will light.

Lamp socket. If a lamp will not light and you are sure the bulb is good, unplug the lamp and test the socket with the continuity tester. After removing the socket (see page 51 on lamp repair), fasten the alligator clip to the metal screw shell and touch the probe to the silver-colored terminal screw. If the socket is good, the bulb will light.

Lamp switch. Unplug the lamp and remove the socket and switch. Clamp the alligator clip to the brass-colored terminal screw and put the probe on the contact tab in the socket. Turn the switch to ON. If the bulb in the tester lights, the switch is good. If the bulb still doesn't light, check to be sure that the contact tab in the socket is high enough to make contact with the bulb. Bend the tab up slightly with your finger or a screwdriver.

Single-pole switch. If you think the electrical problems lie in the switch, first turn off power. Remove the switch from the wall, put the alligator clip on one of the two terminal screws, and touch the probe to the other terminal screw. If the switch is working, the bulb should light when the switch is on and not light when the switch is off.

Three-way switch. A three-way switch, which is used to control light from two different points, has three terminal screws. To test this switch, put the alligator clip on the terminal screw marked "common" (it may be unmarked but of different color). Hold the probe against one terminal screw on the other side and flip the switch. The test light should go on when the switch is in one position and go off when it is in the other position. Now put the probe on the other terminal and repeat the process. The test bulb should go on and off in the opposite sequence.

Four-way switch. A four-way switch, which is rare in most homes, is used in a circuit between two three-way switches and controls power to either of them. To test this switch, attach the alligator clip to any one of the four terminal screws. Then touch the probe to the others in succession. The test bulb should light on only one of them. As an example, fasten the clip to terminal number one (see illustration). If the bulb glows only when you touch terminal number two, it should also glow when you connect three and four. Next flip the switch. The bulb should only glow when you connect one to three and two to four. As a final check of all the switches' ground, put the clip on the metal mounting bracket and touch the probe to each of the terminals. Flip the switch and repeat. The test bulb should not light in any position.

Using a Continuity Tester

Testing a plug-type fuse

Testing a cartridge fuse

Testing a lamp switch

Testing a light switch

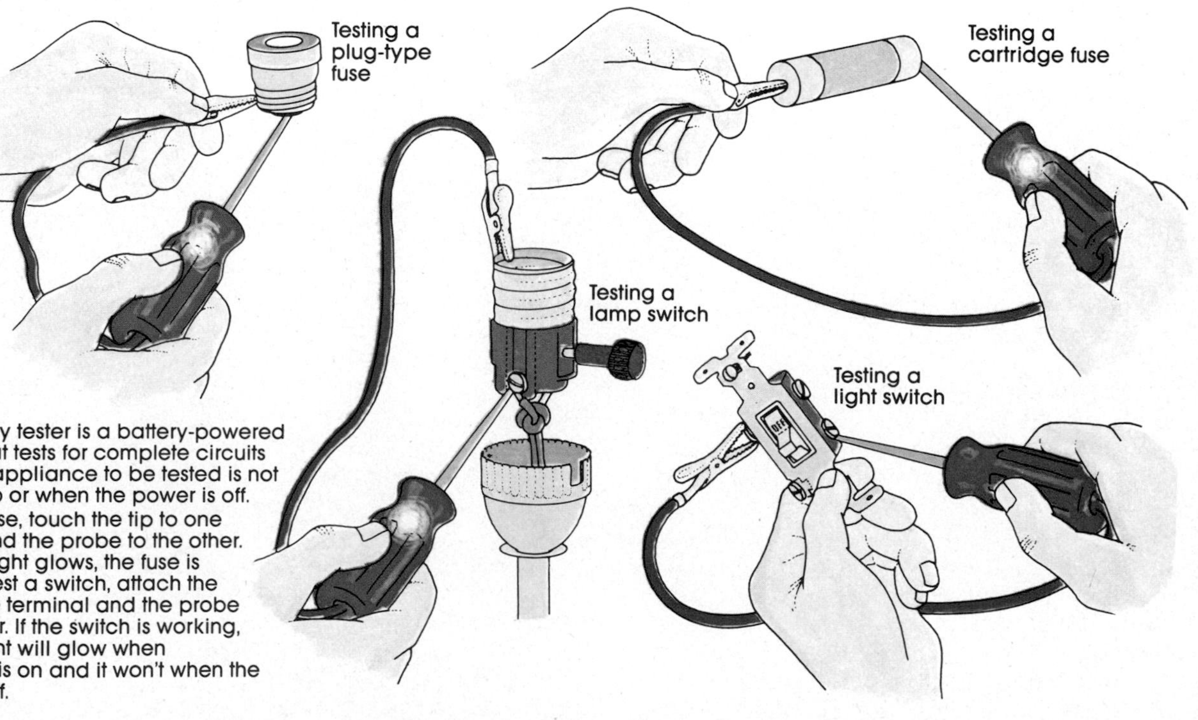

A continuity tester is a battery-powered device that tests for complete circuits when the appliance to be tested is not hooked up or when the power is off. To test a fuse, touch the tip to one contact and the probe to the other. If the test light glows, the fuse is good. To test a switch, attach the clip to one terminal and the probe to the other. If the switch is working, the test light will glow when the switch is on and it won't when the switch is off.

YOUR SAFETY IN MIND

UL Listing

When doing any wiring, even replacing an outlet, make certain that all the materials are approved by Underwriters Laboratories, Inc., an independent, nonprofit testing firm. This is normally indicated by the letters "UL" within a circle stamped on the equipment.

Any manufacturer may send a product to Underwriters, which then runs a series of tests to make sure the product is safe. If so, it is then "listed" by UL. A UL listing means that a product is known to be safe only for its intended purpose. A UL-listed lamp cord, for example, is safe on lamps, but it cannot be used as permanent wiring in the house. Inspectors from the firm periodically visit the manufacturer's plant to see that the product continues to meet UL standards. They may also buy the product at a later date and retest it to see that the safety standards have not slipped.

Electrical equipment without UL listing may be somewhat cheaper, but you may also assume that it is of lower quality. With electrical equipment, this is something definitely to avoid. The slight additional cost of buying UL-listed merchandise is minuscule compared to what it will cost you later to replace cheap equipment when it breaks down.

Look for the Underwriters Laboratories, Inc. stamp on any electrical devices or appliances you buy.

A Word About Codes

When you undertake an electrical project in your home, be sure to do your work according to code. That sounds simple enough, but a problem arises with overlapping codes. These include the National Electric Code (referred to in the rest of this book as the Code or NEC), county codes, and local (city) codes.

Local and county codes basically follow the National Electric Code but are often more stringent. For example, the NEC approves of aluminum wiring, but some local codes now forbid it. To find out how codes apply to your residence, visit your local or county building inspector's office. It is often located in city hall. Some offices have prepared their own list of code requirements, including any variances from the NEC. If they do not have such a paper, ask if they follow the NEC.

Since both the local and county codes are based on the NEC, you should familiarize yourself with it. Don't bother obtaining the Code itself and trying to read it. It is much too complex. Instead, buy a guide to the Code. An excellent example is a paperback that many professional electricians consider their Bible, *Wiring Simplified* by H.P. Richter. Updated annually, it includes practical interpretation of the Code, plus a great deal of information on wiring. You can find it in most large bookstores or write to Park Publishing, Inc., 1999 Shepard Road, St. Paul, MN 55116. In the end, you must abide by the code that has jurisdiction over your residence. If you live in a city, you must follow local codes; if you live in an unincorporated area, follow the county codes.

Basically, you need a permit from your local building inspector's office whenever you add on to your wiring. You do not need one just to replace something. That means you can replace switches and receptacles without a permit, but if you want to extend a circuit, which involves adding more wire, you need a permit.

To do this, first obtain a copy of the local electrical code from your building inspector. Then submit a short written explanation of what you intend to do, complete with a list of the wire sizes you intend to use and details on all other materials, such as fuses, breaker switches, switches, and outlets. Include a simple floor plan or schematic drawing showing where the wire will run and the location of junction boxes, switches, and outlets. It need not be elaborate—just clear. The inspector will then go over your plans and indicate any required changes to meet local codes. Once your plans have been approved, you will receive a permit, for a fee, and when the project is completed, the inspector will check your work to give final approval. Some local codes require that all work be done or approved by a licensed electrician or that the homeowner demonstrate proficiency by taking a test. Check with your building inspector before proceeding.

WORKING WITH WIRE

Wire comes in many types and sizes. You will need to know how to match it to the job, how to strip and splice it. You must also know the basics of grounding.

Most electrical wire is copper, which is flexible and offers little resistance to the flow of electricity. Aluminum wire is sometimes used because it is cheaper, but its use is restricted because of several inherent problems. Aluminum wire offers more resistance than copper wire, so it heats up faster and expands more when electricity runs through it.

The real problem with aluminum wire lies in the connection on switches and receptacles. When aluminum wire is held by a copper or brass screw or spring

◀

From fixing a doorbell to wiring a service entrance box, you may handle many kinds of wire and cable.

connector, the aluminum expands more than the copper or brass when heated. When the power is shut off and the aluminum wire cools, it shrinks. As this process of heating and cooling goes on, the aluminum wire eventually becomes loose in the connection. In time, the connection may overheat and fail. Not only is this annoying, but it can be dangerous. A loose connection causes electricity to spark across the opening, creating a fire hazard.

There are two types of aluminum wire: all-aluminum wire and copper-clad aluminum wire. Many areas now ban the use of all-aluminum wire but permit copper-clad aluminum wire. If you are doing any new wiring in your house, use all copper wire. It is safer, more convenient, and not that much more expensive when dealing with just one house.

Checking for Aluminum Wire

If you think your house may have aluminum wiring, make the following check. Shut off power in a living room circuit. Check that it is off with a voltage tester (see page 11). Remove the wall plate of a receptacle and pull it out. If your house has aluminum wiring, all switches and receptacles will be marked "CO/ALR." Any other markings indicate that the device can be used only with all-copper or copper-clad aluminum wire.

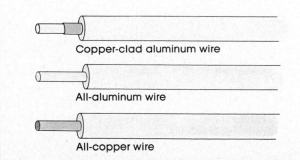

Copper-clad aluminum wire

All-aluminum wire

All-copper wire

Switches and receptacles made before 1971 and rated at 15 or 20 amps were not marked if they were intended for copper wire only. If they were to be used for copper and aluminum, they were marked "CU-AL." After 1971 the Code forbade the use of aluminum wire with these devices. The new devices for use with aluminum wire were then stamped "CO/ALR."

Cu is the chemical shorthand for copper, and *Al* is the abbreviation for aluminum. But to distinguish the change in the types of terminals, Underwriters Laboratories changed the copper listing to *CO* and the aluminum listing to *ALR*.

To be safe, memorize this list:

■ If a 15- or 20-amp device has no markings on it, use only copper wire.
■ If it is marked "CU-AL," use only copper wire.
■ If it is marked "CO/ALR," use copper, copper-clad aluminum, or all-aluminum wire.
■ If it is a push-in terminal device, never use all-aluminum wire. Copper and copper-clad aluminum wire, however, are both acceptable.

WIRE SPECIFICATIONS

Wire Sizes and Use

Electricity is carried by wires that range in size from small lamp cords to high-voltage wires running across the country. The size of wire is important to the efficient flow of electricity. Just as too much water forced through a small hose could cause the hose to burst, too much current (amperes) in a wire could cause the wire to overheat and damage the insulation or even cause a fire. Therefore, the NEC carefully stipulates what type and what size wire can be used in all wiring projects.

Wire sizes are controlled by the American Wire Gauge System, and the size and type are commonly stamped on the insulating cover. The type of wire, such as NM for nonmetallic sheathed cable, is stamped beside the wire size. In wire sizes, the higher the number, the smaller the wire. Thus No. 12 wire is larger than No. 14 wire.

The more current, or amperes, the wire must carry, the larger the wire size must be to facilitate the flow. The illustrations on the right show the wires in actual size, their ampacity (the maximum number of amperes they can carry), and their uses. Note that wires larger than No. 8 and smaller than No. 16 are stranded for flexibility.

The ampere rating given here is for copper wire. If you must use aluminum wire, go two wire sizes higher than the copper rating. Thus, for a 20-amp circuit, you must use No. 10 wire instead of No. 12 copper wire.

Sizes and Types of Wire and Cable

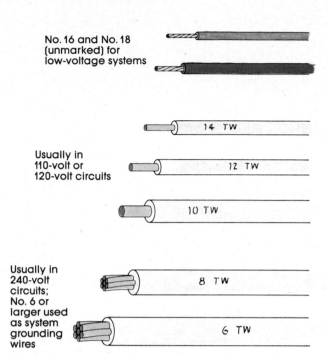

No. 16 and No. 18 (unmarked) for low-voltage systems

Usually in 110-volt or 120-volt circuits

Usually in 240-volt circuits; No. 6 or larger used as system grounding wires

Wire Sizes, Ampacity, and Use		
Number	**Ampacity**	**Use**
No. 4/0	195	
No. 3/0	165	Service entrance wires
No. 2/0	145	
No. 1/0	125	
No. 1	110	Service entrance, ground wires
No. 2	95	
No. 4	70	Individual appliances (240 volts), ground wires
No. 6	55	
No. 8	40	Individual appliances (120 volts)
No. 10	30	
No. 12	20	Small-appliance & lighting circuits
No. 14	15	Lighting circuits
No. 16	10	Doorbells
No. 18	7	Flexible cords, low-voltage systems

Note: Outlets and switches have ampere ratings that are to be matched to the type of wire being used. Most of them are stamped "15 amp," which means they should be used with No. 14 wire. However, the NEC permits you to use 15-amp switches and outlets with No. 12 wire, which has an ampacity of 20 amperes.

Wire Color Codes	
Color	**Function**
Black	Hot wire
Red	Hot wire
Blue	Hot wire
White coded black*	Hot wire*
White	Neutral wire
Green	Grounding wire
Green & yellow	Grounding wire
Bare copper	Grounding wire

*White is always a neutral wire, with the exception of a switch loop (described on page 44), where it must be identified as hot with a dab of black paint or wrapping at the end with black electrician's tape.

Wires are wrapped in colored insulation that designates their function and helps prevent errors in connecting the wires. Neutral wires are white and grounding wires may be green, green and yellow, or left bare. Hot wires are usually black but may also be red or blue. However, never rely solely on the color code for identification. Use a voltage tester to determine if a wire is hot, regardless of its color.

Most house wiring for the lighting circuit uses No. 12 or No. 14 copper wire. If you are installing new wiring, it is advisable throughout the house to use No. 12, which

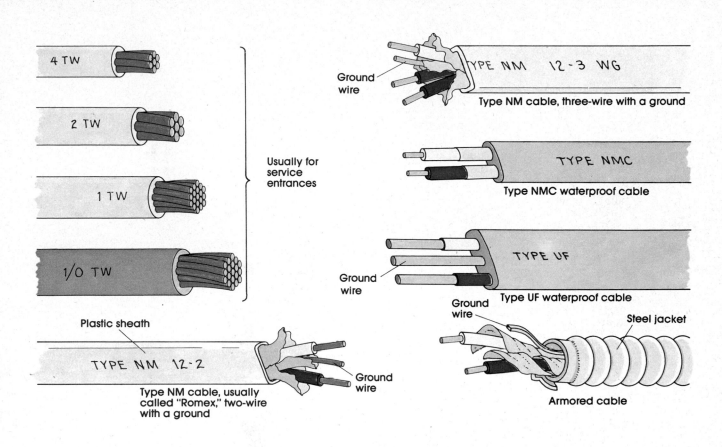

4 TW

2 TW

1 TW

1/0 TW

Usually for service entrances

Plastic sheath

TYPE NM 12-2

Ground wire

Type NM cable, usually called "Romex," two-wire with a ground

Ground wire

TYPE NM 12-3 WG

Type NM cable, three-wire with a ground

TYPE NMC

Type NMC waterproof cable

Ground wire

TYPE UF

Type UF waterproof cable

Ground wire

Steel jacket

Armored cable

is more efficient at carrying current than No. 14. The cost difference is negligible, and many local codes now require No. 12 wire.

When you wire a 240-volt circuit, such as for a kitchen range, use No. 6 or No. 8 wire, depending on what your local codes require. For large appliances that are not 240-volt, such as dishwashers and clothes dryers, use No. 10 wire. No. 12 wire is required when wiring a small-appliance circuit behind the counter in a kitchen. It is also widely used as a lighting-circuit wire, although No. 14 is standard. The small wires, such as No. 16 and No. 18, are used for very low-voltage systems like doorbells. For greater flexibility such as needed in lamp cords, these wires are stranded rather than solid.

Types of Wire and Cable

Technically wire is an individual strand and cable is two or more wires combined in the same sheathing. Each wire in a cable is individually insulated to keep the wires from touching, which would cause a short circuit (see page 22). Some of the most common types of wire are:

Type T wire. This is one of the most common wires in households. The wire is wrapped in thermoplastic (T) insulation as protection against a wide spectrum of temperature differences, from below freezing to the mid–100-degree range.

Type TW wire. This is similar to Type T but has heavier insulation that makes it weather-resistant. It is used for outdoor wiring (although it cannot be buried directly in the ground) and for wiring in damp places such as the basement.

Type THW wire. Similar to Types T and TW, it is more heat-resistant than the others.

NM cable. The most common type of wiring in the house is nonmetallic sheathed cable, widely known by the trade name Romex. It consists of two or more individually insulated Type T wires and a bare copper grounding wire, all coated with a plastic sheath. The space between wires is filled with jute, which can act like a wick to draw in moisture. This wire must not be used where it will be exposed to dampness.

NMC cable. Designed to be used wherever moisture may be present, such as in basements or laundry rooms, NMC cable often has a glass wrapping on each wire. The wires are embedded in a solid plastic sheath to keep out moisture. If NMC is difficult to find in your area, use UF cable, which is waterproof and costs only slightly more.

UF cable. This waterproof cable can be buried directly in the ground. Each of the wires in the cable is embedded directly in plastic that, unless broken, keeps out all water. It is an excellent choice for outdoor wiring or in farm outbuildings subject to moisture.

Armored cable. Usually made of flexible steel or aluminum, the spiral armor itself acts as a partial ground, aided by a strip of aluminum. The wires inside are wrapped in heavy paper to protect them from any abrasive action of the cable. Armored cable is commonly known as BX, a trade name. It cannot be used in damp areas. Some local codes require it in situations where many nails may be driven through the wall. It is sometimes found in older homes but rarely used in modern home wiring.

STRIPPING PROTECTIVE COVERINGS

To expose wires, you must first remove the sheathing, which is usually plastic but may be rubber or may be armored cable. Here's how to remove different types of sheathing.

Plastic or rubber sheathing. Lay the wire flat on a smooth surface. Six inches back from the end, insert the tip of a sharp knife in the sheathing and make a shallow cut down the center to the end. This cut provides a guide for successive passes with the knife. Cut just deep enough to penetrate the sheathing without cutting the insulation around the interior wires. Once the sheathing is split, peel it back and cut it off with a knife or a pair of wire snaps.

Armored cable. Stripping metal requires a hacksaw. Make your cut about 8 inches from the end so you will have plenty of wire for the connections. Hold the saw at right angles to the spiral strip of the armor rather than at right angles to the cable itself. Cut only through the armor; do not cut the aluminum grounding strip or the wires inside. This is not easy and may take some practice. Once you have made a cut through the top of the armor, give the cable a sharp bend at the cut point and twist it back and forth to break the rest of the armor.

Insulation. If you are going to do more than an occasional repair, purchase an electrician's multipurpose tool (see drawing). It makes your work much faster and easier. Simply place the wire in the correct hole (numbered by wire size), squeeze the handles, and rotate the tool to break the insulation. Then pull the insulation off.

Stripping wire with a knife is not recommended because you may partially cut the wire and weaken it. If you have nothing but a knife, don't cut at right angles to the wire; you are more likely to damage the wire that way. Hold the knife blade at about a 60-degree angle, then twirl the wire back and forth until you have cut through the insulation all around the wire. Pull the insulation off with your fingers.

Stripping Cable

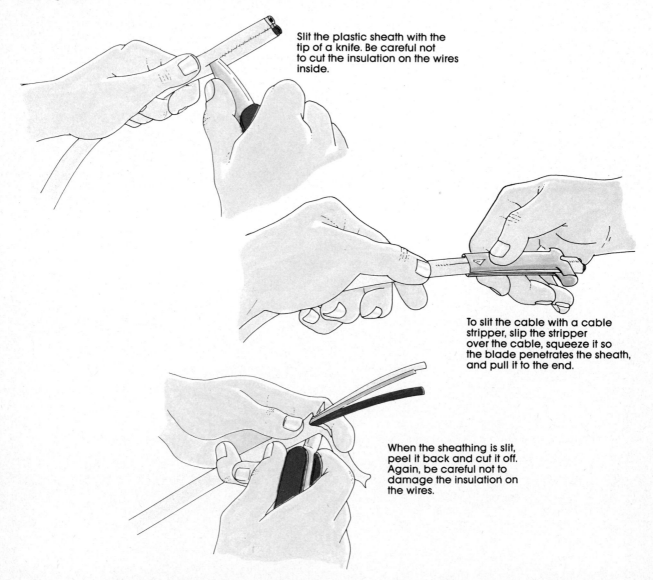

Slit the plastic sheath with the tip of a knife. Be careful not to cut the insulation on the wires inside.

To slit the cable with a cable stripper, slip the stripper over the cable, squeeze it so the blade penetrates the sheath, and pull it to the end.

When the sheathing is slit, peel it back and cut it off. Again, be careful not to damage the insulation on the wires.

Strip armored cable by cutting the spiral metal with a hacksaw. Hold the blade of the saw at a right angle to the spiral and cut only through the top of the armor. Then bend the cable at the cut point and twist to break the rest of the armor.

Stripping Insulation

If you must use a knife, be gentle and hold the knife at a 60° angle rather than perpendicular to the wire. Cut through the insulation without cutting into the wire inside.

With a wire stripper or a multipurpose tool, place the wire in the proper size groove, rotate the tool to cut through the insulation . . .

and then pull the insulation off the end of the wire.

SPLICING WIRE

When wires are spliced, or joined together, the connection must be very tight. A poor connection leads to a drop in voltage, overheating of the wires, and the possibility of a fire from a spark as the electricity tries to jump between loose wires. Wires that are stripped and then spliced together must be fully insulated again. For a tight, safe connection, wires should be spliced with a connector, which may be one of several different types.

Wire nut. The best way to splice wires No. 8 and smaller is with the solderless connector, more commonly called a wire nut. The insulator is a hard plastic shell with a threaded, tapered copper interior. Wire nuts come in different sizes, corresponding to wire sizes. Check the chart on the package to see which size you should use.

To make the splice, first remove the insulation from the end of the wire. Remove only enough so that when

Splicing Wire

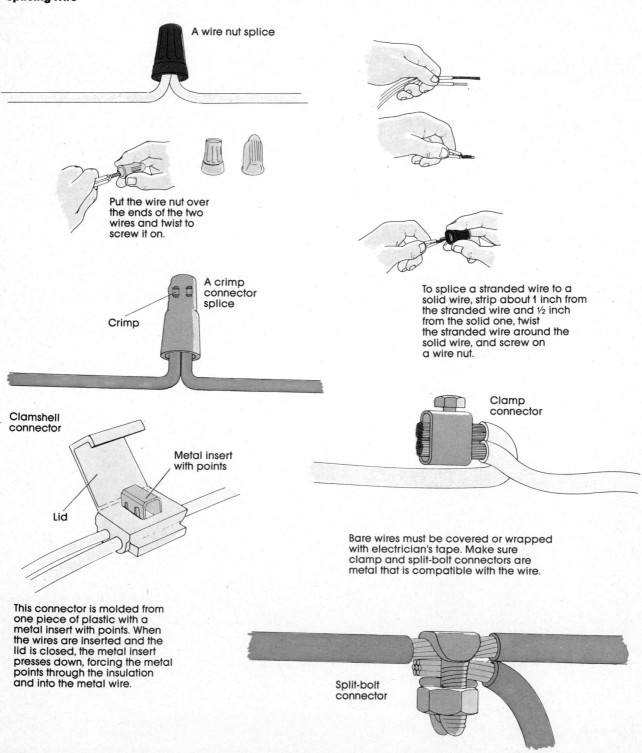

A wire nut splice

Put the wire nut over the ends of the two wires and twist to screw it on.

A crimp connector splice

Crimp

To splice a stranded wire to a solid wire, strip about 1 inch from the stranded wire and ½ inch from the solid one, twist the stranded wire around the solid wire, and screw on a wire nut.

Clamshell connector

Metal insert with points

Lid

Clamp connector

Bare wires must be covered or wrapped with electrician's tape. Make sure clamp and split-bolt connectors are metal that is compatible with the wire.

This connector is molded from one piece of plastic with a metal insert with points. When the wires are inserted and the lid is closed, the metal insert presses down, forcing the metal points through the insulation and into the metal wire.

Split-bolt connector

wires are joined inside the nut no exposed wire will be visible. Twist the two wires together, insert them in the nut, and then screw the nut down as tightly as possible. If some bare wire is still showing, untwist the nut, clip off a little wire at the end, and screw the nut on again. For extra protection against the nut coming loose, wrap the base of the nut and the wires with a couple turns of plastic electrician's tape.

Another type of wire nut has a removable interior with a set screw. You place the wires in the insert, tighten the screw on the wires, and then screw the insulating shell back on the insert.

With the third type, the spring-loaded wire nut, you insert the wires in the connector and then twist the insulator over it. A spring coil inside gives as you turn the insulator around the wire, and then it snaps back to hold the wires firmly. When joining two wires of different sizes, cut the smaller one half again as long as the other, as shown in the drawing. The wire nut will wrap it around the larger wire as you tighten.

Clamshell connector. For No. 12 wires and smaller, you can use a clamshell connector, sized according to the gauge of wire. In this case, do not strip the wires. Place the two ends of wire in the connector and squeeze it shut with a pair of pliers. Tiny teeth in the clamshell puncture the insulation and make the connection from one wire to another.

Crimp connector. Crimps are often used to join No. 16 or 18 wires. First strip the wires and then place them in the crimp. Squeeze the crimp connector together with a pair of pliers or multipurpose tool. A special pair of crimping pliers does a better job if you have a lot of connections to make. Tug lightly on the wires after crimping to make sure they are tight. Do not use crimps with aluminum wire, which expands and contracts too much and eventually works loose.

Clamp connector. You will need a clamp connector to splice larger wires (No. 10, 8, 6, 4, and 2), which are used in running 120 and 240 volts to large appliances such as kitchen ranges or dryers. Strip the two wires to be joined slightly more than the width of the connector, as illustrated. When the bolt at the top is tightened, it forces down a plate that firmly binds the two wires together. Some clamp connectors come with insulated covers that slip over the unit when it is tightened. Others are made of bare copper and must be wrapped in electrician's tape for insulation.

Split-bolt connector. This is one of the easiest connectors to use. As illustrated, slip the split-bolt unit over the continuous wire, put the end of the other wire against the continuous wire, and tighten the bolt. Unless your connector comes with an insulation shield to slip over the entire unit, you must wrap it with electrician's tape.

Caution

If you are splicing aluminum wires, be sure the connector is rated CO/ALR, which means it can be used with either copper or aluminum wire.

All electrical circuits must be grounded. Grounding basically means directing live current into the ground, where there is zero voltage, thus making it harmless. Electricity that is not properly contained in wires and not grounded could pass through you, giving you a severe or even fatal shock.

A key point to remember in grounding is that electricity always follows the path of least resistance. Current flows through a large copper wire more easily than it does through a small one; it flows through copper more easily than it does through most metals; and it flows through copper more easily than it does through you. If you touch something that is electrified but at the same time properly grounded, the current will flow through the grounding wire rather than through you.

Nonmetallic sheathed cable in modern house wiring normally contains three wires: the black one brings the electricity in and is hot, the white one carries the electricity back to the service entrance and is neutral,

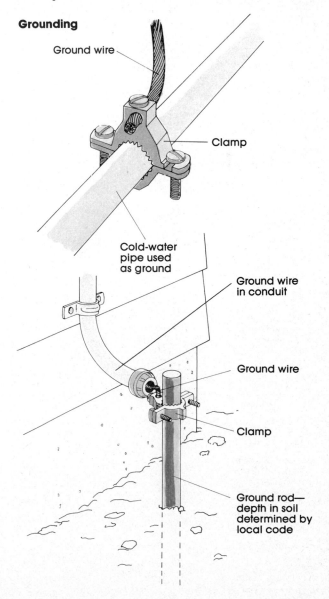

Grounding

Ground wire

Clamp

Cold-water pipe used as ground

Ground wire in conduit

Ground wire

Clamp

Ground rod—depth in soil determined by local code

and the bare copper wire is the grounding wire. It is always connected to a metal switch box in order to ground the box.

For example, say a hot wire had come loose from a terminal screw and was lying against a metal switch box that didn't have a grounding wire. That box would be electrified, and if you touched it, you could get a severe shock. But if a grounding wire were connected to the box, it would pick up that loose electricity and direct it safely to the ground, not through you.

Proper grounding is an essential part of basic wiring, and it's important that you understand these principles. So first consider some of the terminology:

The *ground* is generally a water pipe buried in the ground to carry the returning flow and loose electricity into the earth, where it can no longer be conducted anywhere. This ground, however, is no longer considered sufficient by the NEC, and new houses must have a *grounding electrode system,* as discussed on page 60. A *ground wire* connects the service entrance to the grounding electrode. The term *grounding wire* refers to the bare copper wire in nonmetallic sheathed cable. It does not normally carry current but picks up any abnormal flow of current and directs it into the ground.

Now to complicate things a bit. The white neutral wire in nonmetallic sheathed cable is not entirely neutral. Although normally safe to touch, it carries the same current as the hot black wire on the return trip, thus completing the necessary circle, or circuit. But the white wire acts neutral because it is grounded at the service entrance. Because it must complete a circuit, the neutral wire can never be interrupted by a fuse, circuit breaker, or switch that could interrupt the flow of electricity through it (see page 25). If the neutral wire is not grounded it becomes hot—and dangerous.

White wires are hooked up to receptacles because the current flows uninterrupted through them; white wires are never hooked up to switches, which when off would interrupt the return flow of electricity.

Short Circuits and Ground Faults

A *short circuit* occurs when the flow of electricity takes a short cut instead of following the normal path through the proper wires. This commonly occurs when two wires, one hot and the other neutral, develop worn bare spots and touch each other. This problem occurs most commonly on appliance cords that are dragged about and subjected to considerable wear. Two hot wires touching each other at bare spots also create a short circuit. In a 120-volt circuit, this causes the current to increase, causing an overload and blowing a fuse or tripping a breaker.

In a *ground fault,* a bare hot wire is touching a grounding wire, an armored cable, or a metal conduit. This redirects the current from its normal path toward the ground, interrupting the normal flow. This may also occur when a hot wire rests against the metal covering on a tool or motor.

If the hot wire is touching a ground wire or armored cable, because of the low resistance to electricity, most of the current flows back toward the ground (although

you may feel a shock if you touch the wire). The danger in a ground fault occurs when there is high resistance, such as a hot wire touching the metal frame on a motor or tool that is not properly grounded. That current is then going to travel through your body to the ground.

Grounding and Three-Prong Plugs

If a tool or small appliance has a cord with a three-prong plug on it, the metal frame of the tool or appliance is grounded. If a hot wire were to come loose inside and

Grounded Outlets and Adapters

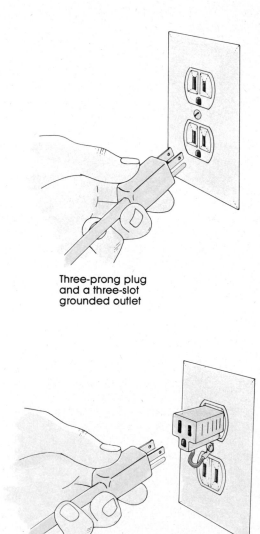

Three-prong plug and a three-slot grounded outlet

Three-prong plug and adapter plug with the pigtail attached to the center screw of a two-slot outlet

touch the metal housing, the current would flow through the grounding wire that is also connected inside to the metal housing, down the grounding wire in the cord, and to the grounding wire hooked to the receptacle. Thus, the current would flow through the grounding wire instead of through your body.

All receptacles in new houses must have three-slot grounding outlets. But many older homes still have only two-slot outlets. That may indicate that there is no grounding wire. If you have an older house with two-slot receptacles, one of the first steps in electrically updating it is to install new three-slot receptacles. But don't do this unless a grounding wire exists. To check this, shut off power on one circuit and remove the faceplate. If there is no bare copper grounding wire attached to the box or switch, the system is not grounded.

When you want to plug into a two-slot outlet, you use an adaptor plug, as shown. Do so only if that receptacle is grounded. Determine this by shutting off the current to that receptacle, removing the faceplate, and checking that a bare copper grounding wire is connected to the receptacle. The purpose of the adaptor plug is to connect the ground wire in the tool cord to the ground in the two-slot receptacle. To do this properly, remove the screw on the receptacle faceplate and then put the screw through the green tab or *pigtail* (short wire) on the plug and back into the receptacle. The ground between the tool cord and the receptacle is now connected. Use the voltage tester to confirm that the screw is grounded (see page 11).

Ground Fault Circuit Interrupters

Using an electrical tool that is properly grounded through a three-prong plug reduces the chance of a serious shock if there is a short circuit or ground fault in the tool. But you might still get a shock, and it could be serious if you are standing on wet grass or damp ground. Water makes it easier for current to pass through you. You might also get a shock if you moved a portable heater with a loose wire around the damp bathroom. This is why *ground fault circuit interrupters* (GFCI) are now required in all bathroom, garage, and outdoor receptacles. They should also be included—but are not yet required—in kitchen, laundry, and workshop circuits. You can even put one in the service entrance for the whole house (see page 58).

Here's how the GFCI works. In standard two-wire wiring (two wires plus a ground), current arrives on the hot black wire and returns on the grounded white wire. The current in both is equal although the white wire will not shock you because it is grounded. If a loose hot wire touches the metal housing on a tool or lawn mower, that current could be redirected through you. The GFCI monitors the equal current in both the black and white wires. If there is any change—namely a drop in the white wire current as it is redirected through your body—the GFCI shuts off all current on that circuit. It does this in about a fortieth of a second, which means the shock is so brief that it will not harm a healthy person.

Three kinds of GFCIs are available. The first type is *built into a receptacle*. You can easily install this one in

GFCI Devices

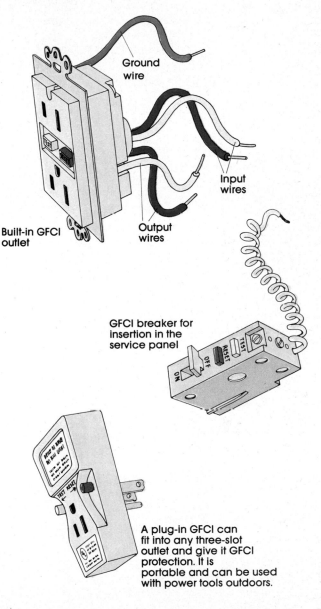

Built-in GFCI outlet — Ground wire — Input wires — Output wires

GFCI breaker for insertion in the service panel

A plug-in GFCI can fit into any three-slot outlet and give it GFCI protection. It is portable and can be used with power tools outdoors.

the bathroom, garage, or outdoor receptacle. This type, if installed on the first receptacle on a circuit, will also protect against any ground faults that might occur in tools plugged into other receptacles farther along on that same circuit. It is the easiest to install. The second type is *installed directly in the breaker panel.* (See page 31 for how to install a GFCI.) It combines a breaker switch in 15, 20, 25, or 30 amps with the GFCI. It is available for 120-volt two-wire or three-wire circuits and for 240-volt circuits. This combination protects against overloads and ground faults. The third type can be *plugged into an existing receptacle* for temporary use. Either this type or a GFCI installed in a receptacle must be used at construction sites where permanent wiring has not yet been installed in the building. In these cases, electricity comes from a temporary power pole.

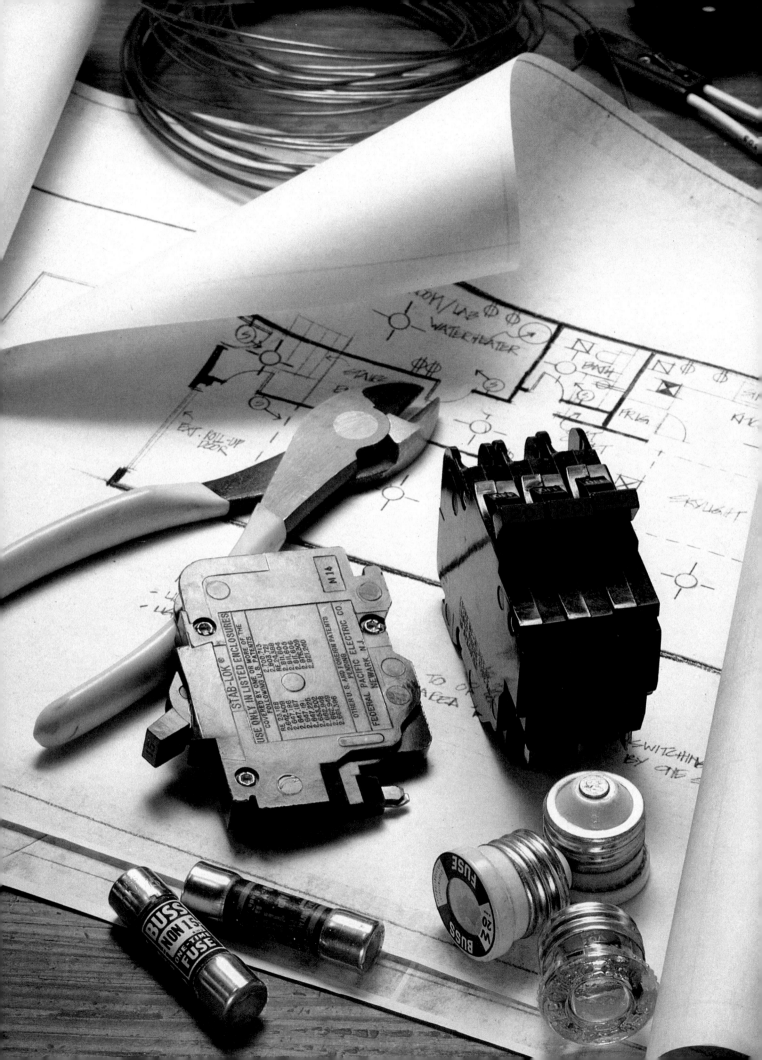

CIRCUITS, FUSES, & CIRCUIT BREAKERS

Well-planned circuits will give you the power you need, when and where you need it, without an overload. You can install different kinds of fuses and breakers, from time-delay fuses to GFCIs.

For electricity to do work, it must complete a circle. From this concept we get the word circuit. Electricity flows along one wire from the source, through the light bulb, and then along a different wire back to the source.

From the fuse or breaker box, wires run to the individual circuits in your house, some for the living room, some for the kitchen, some for the baths, and so on. In each standard household circuit, electricity travels along the black hot wire to an electrical device and returns on the grounded white neutral wire to a ground wire at the service entrance.

If too many electrical appliances are plugged into one of these circuits, it begins drawing more electricity than the circuit wire is capable of handling. The circuit becomes overloaded. When this happens, a fuse is blown or a breaker is tripped. These are known as overcurrent devices, which prevent the circuit from taking on dangerous amounts of electricity. Without that overcurrent device, the wire would grow increasingly hot from the excess electricity. The protective insulation would melt, and a fire might result.

Again, this is why you must never use a larger fuse than the circuit wire size can handle. Putting a 30-amp fuse in a 15-amp circuit allows you to draw much more electricity without blowing the fuse, which dangerously overloads that circuit (see page 28).

Circuits

From the fuse box or breaker panel, individual *branch* circuits run through the house. Although it is possible to have all the lights and appliances in the house on one wire, it is highly impractical. First, it would have to be a very large wire to handle the electrical demand. Second, if the fuse or breaker for the single circuit did go off,

◀

Using the right kinds of fuses and breakers will protect your home from a circuit overload.

everything in the house would be shut down. With individual circuits for different parts of your house, only the part of the house containing that circuit is affected when something goes wrong. Having more circuits also means less voltage drag (power loss). One circuit wire covering an entire floor would, because of the lengthy wire needed, have more resistance than two shorter circuits covering the same area. There are three basic types of branch circuits for a residence: the lighting (or general-purpose) circuit, small-appliance circuit, and individual-appliance circuit.

Lighting Circuit

The lighting circuit basically accommodates switches and receptacles that run through the house for the lights, radios, televisions, portable electric heaters, clocks, and similar items. It does not include kitchen appliances.

In planning to install a lighting circuit in a new house, or to add a circuit when remodeling, first ask the question: How many circuits do I need? According to the Code, you need one circuit for every 600 square feet of floor space in the house or three watts of power for every square foot of usable space in the house. But professional electricians suggest that you should have one circuit for every 500 square feet. This may mean adding one or two circuits in a new house or large addition, but the extra outlets and power will be worth it to you in the long run. And if you are even remotely considering some day turning the garage into a family room or putting a small bedroom up in the attic, then include that area in your floor space figures. A properly wired house should never have a blown fuse or a tripped breaker. In making your calculations, consider the following example:

If you have a two-story house that is 24 by 36 feet, then each floor contains 864 square feet, for a total of 1,728 square feet. Divide that amount by 600 (one circuit for every 600 square feet), and you find that you need a minimum of 2.88, or three, circuits according to the Code. But using one circuit for every 500 square feet, the answer is 3.45, which means you need four circuits. If you use No. 14 wire, which has an ampacity of 15 amps, one circuit can handle 1,800 watts (15 amps × 120 volts). You might be able to quickly estimate how many watts might be on that circuit at a time, such as in the evening with the television on, several lamps lighted, and the vacuum cleaner in use.

With modern appliances abounding and pulling larger and larger wattages, some professionals recommend all new wiring be done with No. 12 wire, which has an ampacity of 20 amps. This allows you to run up to 2,400 watts (20 amps × 120 volts) on that circuit without risking an overload.

If you have an area such as a garage that you might later convert to a living space, one easy way to approach the problem is to run a separate circuit to it with just one switch and an overhead light. Later, when you are ready to refinish the area, you can install the additional receptacles and switches and tie into the existing circuit there.

Let's say the garage is 20 × 20 feet, or a total of 400 square feet. You must allow 3 watts for every square foot, which means you need 1,200 watts for that room. Since No. 14 wire is rated at an ampacity of 15 amps (see page 16), it will handle 1,800 watts, which covers the 1,200.

Small-Appliance Circuit

To many people, the kitchen is the heart of the house. It is certainly the heaviest user of electricity—you may find a toaster, a coffee pot, an electric frying pan, a blender, and other such items all going at once. Someone may be ironing there, too. Some of these appliances consume 1,000 watts or more.

Because of this heavy demand for electricity, the Code requires that new kitchens have a minimum of two 20-amp circuits just for the small appliances. These must be separate from the lighting circuit in the kitchen. Each of these small-appliance circuits must be wired with No. 12 wire, which has an ampacity rating of 20 amps. Each circuit will handle 2,400 watts (20 amps × 120 volts), which is not all that much, considering that an electric toaster may consume 1,500 watts by itself.

Circuit Plan for a Typical Home

Lighting and Outlet Circuits

— Kitchen light, porch light, family room, and upstairs bath

— Small appliance circuit

— Small appliance circuit

— Dining room, downstairs hall, and two upstairs bedrooms

— Upstairs hall, linen closet, downstairs bath, entry hall light and outlets, and porch light

— Living room and master bedroom

— Garage, workshop, and outside light

Individual Appliance Circuits

— Electric clothes dryer

— Range and oven

— Air conditioner

— Dishwasher

— Garbage disposer

⊙ Light

s Switch

● Outlet

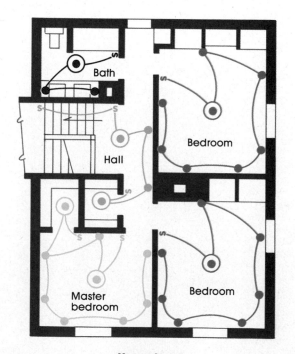

Upper Level

When you wire the receptacles on these two circuits, put every other receptacle on the first circuit and the intervening ones on the second circuit. This way, if several individual appliances are plugged in next to each other on a counter, they will not be operating on the same circuit, which could overload it.

Because you will be using small appliances in other rooms, for such things as an extra refrigerator in the pantry or a coffeepot and blender in the dining room, the small-appliance circuit must extend to those rooms as well. Again, it cannot be tied to the lighting circuit. In addition, a separate 20-amp circuit must run to the laundry room for the washer and iron.

Individual-Appliance Circuit
Certain large appliances, such as kitchen ranges, water heaters, and built-in ovens, are normally wired on their own individual circuits. If you are planning an addition, you will need an individual circuit for any of the following items:
- Kitchen range
- Separate built-in oven
- Dishwasher
- Garbage disposal
- Compactor
- Water heater
- Clothes dryer
- Oil-burning furnace motor
- Pump on a well
- Permanent appliance rated at more than 1,000 watts, such as bathroom heater

These appliances may require a circuit of 120 volts, 240 volts, or a combination of both, depending on their demand for current.

In summary, you must plan on three basic types of branch circuits if you are building a new house or doing some extensive remodeling and rewiring. You need one 15-amp lighting circuit for each 600 square feet of flooring (one 20-amp circuit for each 500 square feet is better), two 20-amp circuits in the kitchen for the individual small appliances, a separate 20-amp circuit for the laundry room, and individual circuits for major appliances.

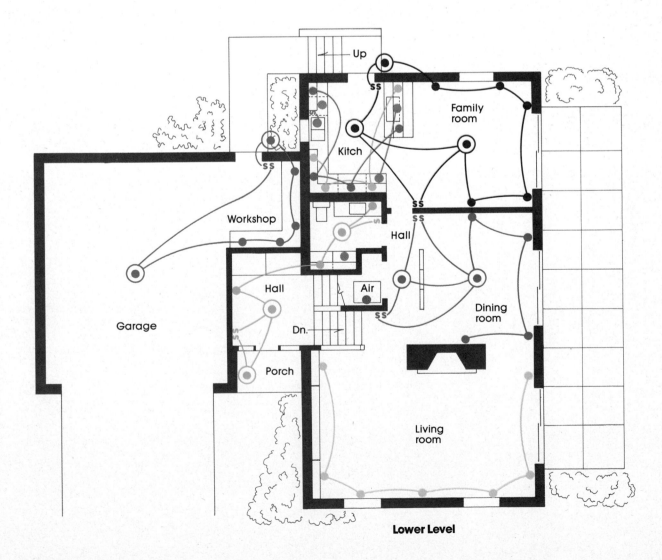

Lower Level

FUSES

A fuse is a thin piece of metal that is designed to melt immediately, or blow, if too much current flows through it. The fuse is totally enclosed to prevent the sparks and hot metal from flying about when this happens. When the strip of metal in the fuse melts, it creates a gap that the electricity cannot bridge. Power in that circuit is cut off, thus protecting it from further abuse.

When a Circuit Overloads

A fuse blows when too many amps flow through it, such as when too many appliances are plugged in at once, all demanding more electricity than the wire can carry. It will blow in a short circuit, which also greatly increases the amount of electricity in the wires. In addition, a fuse will tend to blow if it is loose in its socket. This is something to check if a fuse blows for no apparent reason.

Before replacing a fuse or resetting a breaker, always find out what caused the overload so it won't be repeated. You can guess quite accurately what the problem is by looking at the fuse top. If the glass is black and you cannot see the metal strip inside, it was probably a short circuit. This causes the fuse to blow instantly under high heat. If the glass is clear and you can see the metal strip, it indicates a circuit overload. This causes the strip to heat up steadily until it finally melts.

If a fuse blows almost immediately after you plug something in, regardless of what circuit you try, that indicates a short circuit somewhere in the appliance itself. The problem is commonly in the cord, where two wires have been exposed and are touching each other. Check near the plug and where the wire enters the appliance. Make sure the appliance is unplugged while you check or correct the problem.

If a fuse routinely blows on one circuit, it is most likely a simple overload. You will just have to use fewer lights and appliances on that circuit.

If a fuse blows when you start an electric motor in your shop, you might be able to correct the problem by using a time-delay fuse. This handles the temporary overload required to start the motor. But if the problem persists, then suspect a short circuit in the motor. Check the cord first.

If your house is protected by cartridge fuses, you cannot tell by looking at them if they are blown or not. To check, cut all power to the house, remove the suspected cartridge, and check with a continuity tester, as described on page 12. After correcting the overload problem, install the new cartridge and turn the main power back on.

Types of Fuses

Fuses are directly related to the type of wire in the circuit they are protecting. It's important to remember that wire has a certain ampacity, or maximum amount of current it can carry without the risk of overheating (see page 16). For example, No. 14 copper wire, commonly used for lighting circuits, has an ampacity of 15 amps: use nothing higher than a 15-amp fuse or breaker. Here are the most common types of fuses:

Plug or Edison-base fuse. One of the most common types of fuses, particularly on older houses, is the plug or Edison-base type. It is rated only up to 30 amps. The amp rating, stamped on the metal fuse link, is visible through the protective glass cap. The plug fuse has a ceramic base that screws into the socket in the service entrance. Because Edison-base fuses will screw into any socket in the fuse box regardless of the actual amp rating, your circuits could have fuses that allow circuit overloads. When moving into an older house, check that the previous occupants have not put in larger-amp fuses than permitted in order to keep from blowing fuses. It is a dangerous practice.

Types of Fuses

Plug or Edison-base fuse. The metal strip that shows through the glass completes the circuit. A break in the strip indicates an overload; blackened glass usually indicates a short circuit.

Time-delay fuse. The metal strip is spring-loaded to allow a temporary overload, as when a motor is starting.

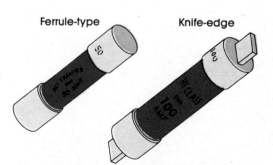

Cartridge fuses. These fuses give no outward indication when they have blown — you must replace them or use a circuit tester to find out. The ferrule-type fuses go up to 60 amps. Fuses over 60 amps have knife-edge contacts.

Fuses for Safety

Nontamperable or type-S fuse. The fuse will not fit a standard fuse socket. The adapter is put in the socket and cannot be removed—then only the properly rated fuse will fit into the adapter.

Nontamperable or Type-S fuse. The nontamperable fuse, also called the Type-S fuse or Fusestat, is designed to prevent homeowners from using fuses with higher amp ratings than the circuit will safely bear. The top of the fuse looks quite similar to the plug-type fuse, with the fuse link visible through the protective glass cover. But the base of the Type-S fuse will not fit in the standard fuse socket. Instead, the Type-S fuse screws into an adaptor that is fitted into the fuse socket. A short piece of spring steel protrudes from the side, allowing you to screw the adaptor into the socket but preventing you from unscrewing it. Thus, you can use only the properly rated Type-S fuse in that circuit. If a new fuse box is being installed, the Code requires it to take nontamperable fuses. In many areas, the power company will require you to install them if they think you are using fuses with higher ampacity than is safe for that circuit.

Time-delay fuse. Time-delay fuses are useful in situations where short and temporary overloads exist on the circuit. A common example is using electric motors in the workshop, where a high amount of amps may be used in starting the motor. A fuse rated at 15 amps will carry that amount indefinitely but will blow if the circuit exceeds it. But a 9-amp motor, for instance, may demand nearly 30 amps for a second or two when starting. The time-delay fuse will carry that overload for several seconds while the electric motor is demanding a great amount of current to get started. If the demand runs too long, the time-delay fuse will then blow, protecting the circuit and the motor.

Cartridge fuses. Cartridge fuses are most commonly used as the main power fuse. In many cases, two cartridge fuses are attached behind a removable plate in the entrance box. Pull out that plate to remove the fuses and shut off all power to the house. In some older houses, cartridge fuses are also used to protect the branch circuits. There are two basic types of cartridge fuses. The *ferrule-type cartridge* has metal caps at each end and is rated from 10 to 60 amps. It is normally used to protect large individual appliances in the kitchen or shop. The *knife-blade cartridge* has metal blades at each end that fit into spring clips in the fuse box. It is designed for more than 60 amps and is often used to protect the entire service entrance.

CIRCUIT BREAKERS

Circuit breakers are used on virtually all new homes being built in this country, and have been since the early 1960s. The chief advantage is that they do not have to be replaced like a blown fuse. Instead, an overload causes the breaker, which resembles a wall switch, to flip from ON to OFF. Once you have corrected the overload problem, simply reset the breaker. To do this, first push the switch as far as it will go beyond the OFF setting and then flip it to ON. On some models, you need only flip the switch back to ON to reset it.

Another advantage of the breaker is that it acts like a time-delay fuse. It will carry small overloads for up to 30 seconds and large overloads, as in starting large electrical motors, for several seconds before tripping.

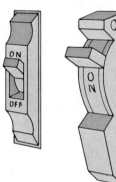

Three styles of circuit breakers by different manufacturers. The double breaker (at right) has a bar connecting the two handles and is used to protect 240-volt circuits.

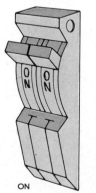
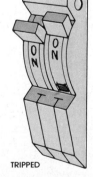
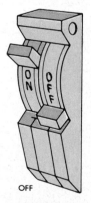

ON   TRIPPED   OFF

Some breakers go to the OFF position when tripped. Some go halfway to OFF or (as shown above) to a TRIPPED position. These must be switched to OFF then ON to restore service.

You can replace your screw-in fuses with this type of breaker (at the same rating). An overload will cause the button to pop out, exposing a colored band. To restore service, just push in the button.

Replacing a Breaker Switch

If, after a circuit breaker trips and has been properly reset, there is still no power on that circuit, the problem is probably a defective breaker switch. To check, first throw the main power switch at the top of the panel to OFF. This stops all power to the service entrance. Next remove the panel covering the breaker switches. As an added precaution, stand on a couple of dry boards, such as 2 × 4s. Turn the power back on and then check the suspected breaker with a voltage tester (see page 11). Put one prong on the terminal screw and the other on a ground, such as the panel. If the tester fails to light, the breaker is not functioning.

To remove the breaker, turn the main power switch to OFF. Grasp the breaker between your thumb and forefinger and pull on the end with the power load screw. When you have removed the breaker, unscrew the black load wire. If you are working with a breaker near the top of the panel, be very careful not to touch the wires above the main power switch. These wires are still carrying electricity.

If you do not already have a few replacement breakers on hand, look carefully at the type you have. Different brands are of different styles, which are not all interchangeable. Take the nonfunctioning one with you to the hardware store to get an exact match.

To install the new one, first turn the main power switch to OFF. Next loosen the screw at the load end of the breaker, insert the wire, and tighten the screw. As illustrated, hold the breaker at a 45-degree angle and hook the end nearest the load wire first. Then push the other end down so the contact points lock over the service-entrance prong. Replace the panel over the breakers and flip the main power switch to ON.

Removing a Breaker

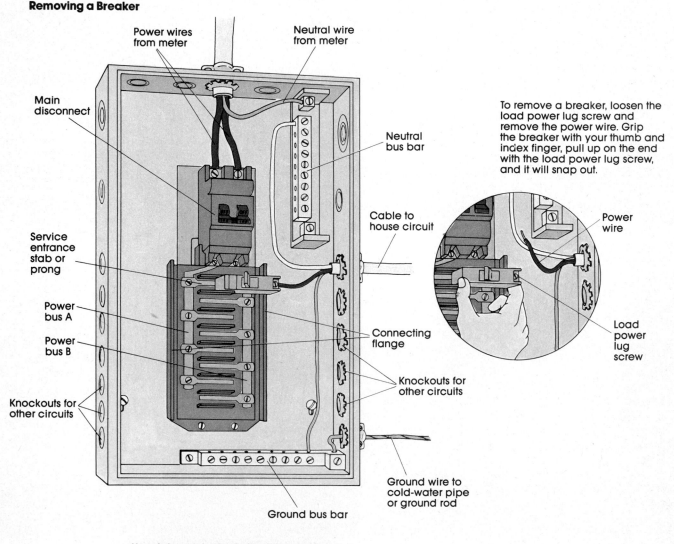

Power wires from meter

Neutral wire from meter

Main disconnect

Neutral bus bar

Service entrance stab or prong

Cable to house circuit

Power bus A

Power bus B

Connecting flange

Knockouts for other circuits

Knockouts for other circuits

Ground bus bar

To remove a breaker, loosen the load power lug screw and remove the power wire. Grip the breaker with your thumb and index finger, pull up on the end with the load power lug screw, and it will snap out.

Power wire

Load power lug screw

Ground wire to cold-water pipe or ground rod

Here is how a typical circuit is connected to the breaker and the panel. For clarity, we show only one circuit. The breaker has contacts that connect it to one of the power buses and the neutral bar when it is snapped into place.

Installing a GFCI Breaker

A GFCI breaker is similar to a regular breaker except that in addition to the ON–OFF switch and a power lug, it has a test button, a reset button, a neutral lug, and a neutral pigtail wire.

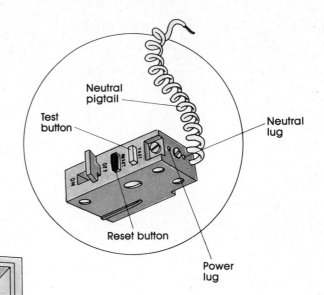

Neutral pigtail

Test button

Neutral lug

Reset button

Power lug

A GFCI breaker installs just like a regular breaker except that the neutral wire from the circuit cable connects to the neutral lug of the breaker, and the neutral pigtail on the breaker connects to the neutral bus bar in the panel.

Installing a GFCI

Ground fault circuit interrupters (GFCIs) can be life-saving devices. (See page 23 for discussion of ground faults and different types of GFCI devices.) New houses being constructed must have GFCIs for the bathroom, garage, outdoor outlets, and swimming pool or hot tub. It's also wise to use them wherever there might be both water and electricity, which includes the kitchen and laundry room.

Note that a GFCI must have its own individual circuit. It cannot be on a circuit that has a neutral wire sharing another circuit. This may cause the GFCI to trip repeatedly. To install a GFCI in place of a standard breaker, follow these steps:

1. Turn off the main power switch and remove the panel over the breaker switches.
2. Remove the black load wire from the selected circuit

breaker and remove that breaker by pulling on the end opposite the load wire.
3. Connect the neutral pigtail wire that comes with the GFCI to the neutral bus bar (see illustration).
4. Connect the white neutral wire of the circuit to be protected to the terminal on the breaker marked "load neutral." Connect the black hot wire of that circuit to the terminal marked "load power."
5. Put the GFCI in place as illustrated for a standard breaker.
6. Replace the panel over the breaker switches.
7. Make sure the GFCI is still in the OFF position, then turn the main power back on.
8. Turn the GFCI to ON.
9. Testing instructions come with each type of GFCI. They all have a test button for this purpose. Follow the instructions on your model carefully.

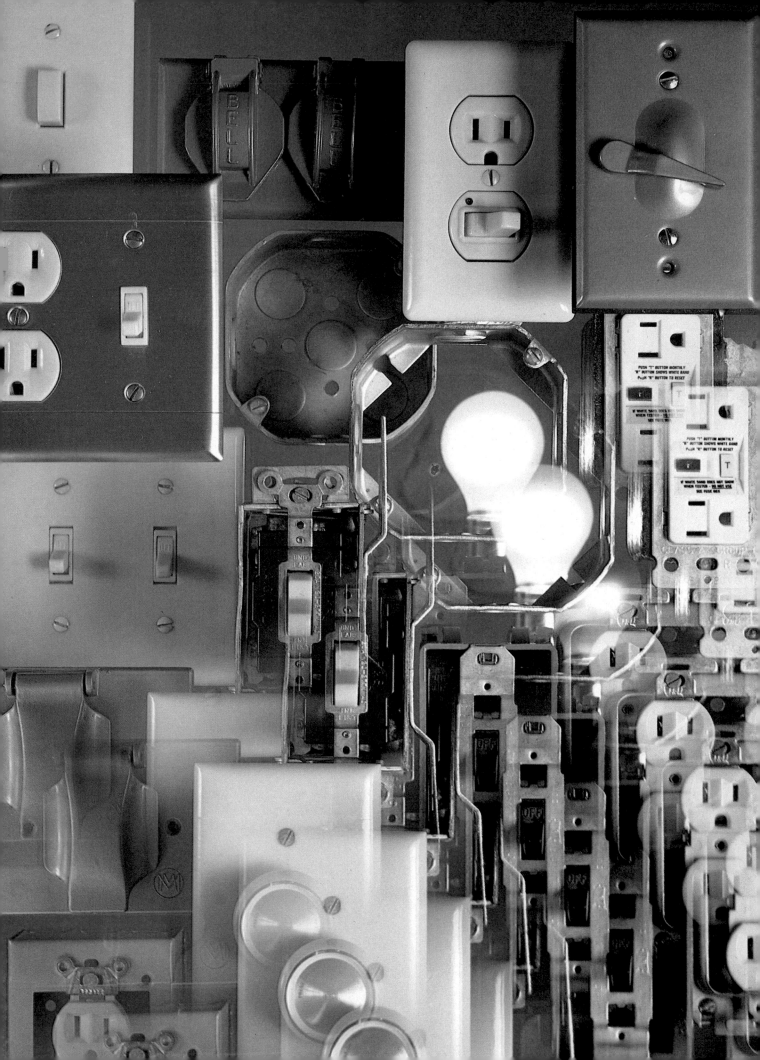

BOXES, SWITCHES, & RECEPTACLES

These simple devices provide access to the circuits in your walls. Whether you are installing new switches and outlets or replacing old ones, you may want to consider some special types.

One of the most common tasks for the home electrician involves the installation of switch boxes, or utility boxes, and the switches and receptacles for those boxes. Although there are several different types of switch boxes, they are all constructed to hold either a switch or a receptacle, or a combination of both. And if you need more than one box in the same location, you can put two or more together, a process called *ganging* boxes.

The sizes of boxes, switches, and receptacles are standardized so no matter what you buy, they all go together. Those bearing a UL stamp of approval may be more expensive, but you can be assured they are safe. As with all electrical equipment, look for the sales and then buy everything at once. In a medium to large project, it can amount to a considerable savings. The following survey will help acquaint you with the sometimes confusing variety of boxes, switches, and receptacles available.

Boxes

Wherever you cut branch circuit wiring in order to splice it into another wire or attach it to a switch or outlet, those exposed ends must be contained in a junction box, switch box, or ceiling box. In addition to housing the switch or receptacle, the box keeps the exposed wire ends away from flammable material. All boxes must be covered and at the same time always accessible. In some cases, such as junction boxes, the cover is solid. Other covers are designed for outlets or switches. Boxes are made either of plastic, or steel with a galvanized finish. Fiberglas boxes are also available in some areas.

Plastic Boxes

Plastic boxes are cheaper than metal and are fine for a basic lighting circuit in a new house. They are made from a hard thermoplastic intended primarily for use with nonmetallic sheathed cable. Although the Code says that the wiring does not have to be clamped in the box if it is supported within 8 inches of the box by a staple, some local codes require clamps on the box. These clamps screw together as shown, one inside the box and one outside, to hold the wire in place.

Plastic boxes normally come with two 16d nails fixed in the mounting bracket. These nails are driven into the stud or ceiling joist, depending on the use of the box. Several different types of plastic boxes are available. You can use them in new installations and in modernizing old work.

Because plastic boxes do not conduct electricity, they do not need to be attached to the grounding wire in the cable. Instead, attach the grounding wire to the green hexagonal grounding terminal on the switch or receptacle.

Types of Plastic Boxes

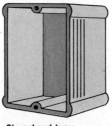

Standard box

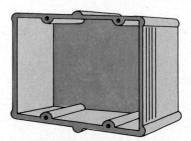

Gang box for two switches or receptacles

◀

Switches, boxes, and even outlets come in many shapes and combinations to serve different needs.

Metal Boxes

A variety of metal boxes are available in different sizes and shapes, designed for different uses (see illustrations). Metal boxes have a threaded hole at the rear so that the grounding wire can be attached to the box with a sheet metal screw. Basic wiring is done with a box 2½ inches deep so that you have room to work and the box is not overly crowded with wires. But metal boxes range in depth from ½ inch to 3½ inches.

For general lighting circuits, most metal boxes have a pair of clamps inside to hold the nonmetallic wires. To use these clamps, first remove the knockout (see the next section), then slip the wire under the clamp. Tighten the screw to force the clamp against the wire.

Two boxes can be joined together, or ganged, so that four devices can be installed in the box. The sides of this type of box are held in place by a screw. If you need only two devices in a box, such as a switch and an outlet, 4-inch-square boxes are made specifically for this purpose.

Types of Metal Boxes

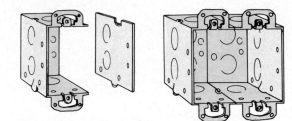

Standard metal box capable of ganging

Utility box

Drywall box

Outdoor box

Clamping Wire in Boxes

When wires are fed through knockout holes in boxes, they should be clamped to hold them securely in place. Here are three basic types of clamps for that purpose:

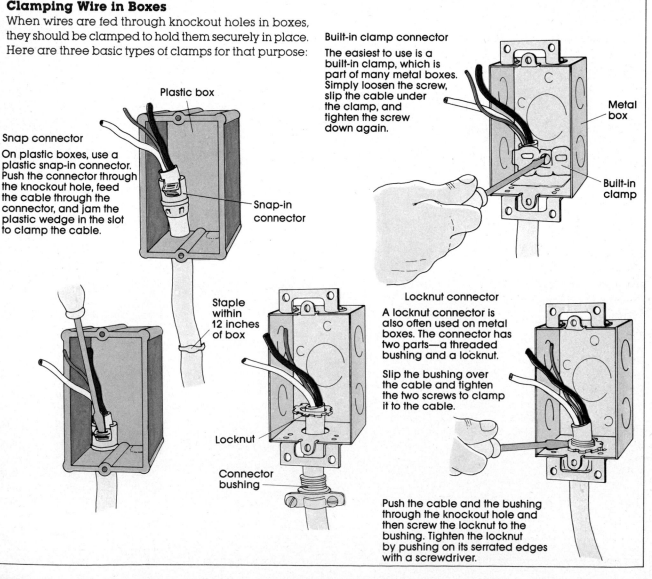

Snap connector

On plastic boxes, use a plastic snap-in connector. Push the connector through the knockout hole, feed the cable through the connector, and jam the plastic wedge in the slot to clamp the cable.

Plastic box

Snap-in connector

Staple within 12 inches of box

Locknut

Connector bushing

Built-in clamp connector

The easiest to use is a built-in clamp, which is part of many metal boxes. Simply loosen the screw, slip the cable under the clamp, and tighten the screw down again.

Metal box

Built-in clamp

Locknut connector

A locknut connector is also often used on metal boxes. The connector has two parts—a threaded bushing and a locknut.

Slip the bushing over the cable and tighten the two screws to clamp it to the cable.

Push the cable and the bushing through the knockout hole and then screw the locknut to the bushing. Tighten the locknut by pushing on its serrated edges with a screwdriver.

Knockouts

All boxes come with *knockouts,* which are sections that can be removed from the front, side, top, or bottom, depending on where the wires are running. In plastic boxes, the knockouts are just a thin section in the box wall that is punched out with the handle of a pair of pliers. On metal boxes, all but a small part of the knockout has been already cut out. To remove it, just rap the knockout sharply with the plier handle, then twist it back and forth to break off the "hinge." Some of the knockouts on metal boxes have small slots through which you can insert a screwdriver to pry the knockout loose.

You must not remove any more knockouts than you need. If you remove one and then decide to go elsewhere with the wires, the Code requires you to seal the wrong hole with a knockout closure. Two types of closures are available in most hardware or electrical stores. For general work, use the metal disc with tension clips around the edge. Simply press it into the unused hole. For larger openings, there are two discs, one on each side of the hole, which are held together by a screw through their center.

Number of Wires Permitted in a Box

The Code limits the number of wires that can be contained in any given switch box, junction box, or ceiling box. In addition to the wires, these regulations include other items often found in boxes. All are referred to as *conductors.* The rules are based on the size of the wires used and the cubic inch size of the box. They are primarily designed to prevent you from overcrowding the box, which, apart from making it difficult to work, might damage the wires. If this matter seems somewhat confusing, just use common sense and do not crowd the box.

The chart below may require some interpretation:

1. If a wire originates and ends within the box, such as the grounding jumper wire from the receptacle to the metal box, do not count it as a conductor.

2. The wires from a fixture to the wires in the box are not counted as conductors.

3. Count as only one conductor the two cable clamps often built into metal boxes.

4. A bare grounding wire entering and leaving the box is counted as only one conductor, regardless of the number of such wires.

5. Deduct one from the numbers on the chart for each hickey, fixture stud, or mounting strap (see page 52 on replacing ceiling fixtures) in the box.

6. Do not count a switch or receptacle in the box.

As an example of the above, say you had a standard metal switch box 3 inches high, 2 inches wide and 2½ inches deep (3 × 2 × 2½), with two built-in cable clamps. According to the Code, you are permitted 6 conductors in the box when using No. 14 wire. This would include two incoming conductors (hot and neutral) plus a bare grounding wire. For a middle-of-the-run connection, there would also be two outgoing conductors plus a bare grounding wire. That accounts for five conductors, since the bare grounding wire is counted only once. The two cable clamps in the box are counted together as one

Knockout Closures

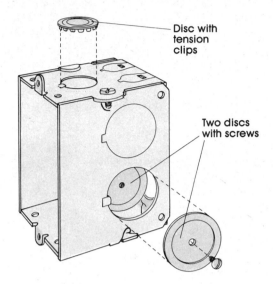

Disc with tension clips

Two discs with screws

more conductor. The total is now six, which is the maximum. You can still install the switch or receptacle, since they are not counted.

But if you were using No. 12 wire, the limit would be five conductors. Technically, you should therefore use a larger box or the same size without cable clamps to remain under the maximum. If you have any doubts, talk it over first with your local inspector.

Number of Conductors Permitted in a Box

Box size	Maximum number of conductors			
	No. 14	No. 12	No. 10	No. 8
Round or Octagonal				
4 × 1¼	6	5	5	4
4 × 1½	7	6	6	5
4 × 2⅛	10	9	8	7
Square				
4 × 1¼	9	8	7	6
4 × 1½	10	9	8	7
4 × 2⅛	15	13	12	10
Switch Boxes				
3 × 2 × 1½	3	3	3	2
3 × 2 × 2	5	4	4	3
3 × 2 × 2¼	5	4	4	3
3 × 2 × 2½	6	5	5	4
3 × 2 × 2¾	7	6	5	4
3 × 2 × 3½	9	8	7	6
Junction Boxes				
4 × 2⅛ × 1½	5	4	4	3
4 × 2⅛ × 1⅞	6	5	5	4
4 × 2⅛ × 2⅛	7	6	5	4

SWITCHES

Four basic types of switches are used in houses. While not all four may be present in your house, two or three of them probably are. You need to know how all four work in case you do have them or decide to install them. If a switch goes out, make sure that you replace it with another of the same type. In other words, replace a single-pole switch with a single-pole switch, not with a four-way switch. That is the only way it will work properly. If you are installing wiring in a new house or ripping out all your old wiring, then consider using any of these switches where they would prove most useful.

Single-pole switch. This is the most common switch in any house. It is used primarily to turn on a ceiling light that is operated by only one switch or to open a circuit where a lamp is plugged in. The handle, or toggle, is marked with ON and OFF lettering. The switch has two brass-colored screws or two holes to insert copper wire.

Three-way switch. Contrary to what the name implies, the three-way switch is used to control a light or receptacle from two different points, not three. But you can always tell the three-way switch by its three terminal screws and its toggle with no ON or OFF lettering. Two of the screws are either brass- or silver-colored, and the third is black or copper-colored. Three-way switches are commonly used at the top and bottom of stairs or at different ends of a hallway or living room. They are always used in pairs. Connecting them also requires special three-wire cable (see page 17).

Four-way switch. The four-way switch is used in conjunction with three-way switches to control lights from three or more locations, such as in workshops or large living rooms. You can easily recognize the switch by its four terminal screws, all brass-colored. It has no ON or OFF lettering on the toggle.

Double-pole switch. This switch looks similar to the four-way switch in that it has four brass-colored screws. The key difference is that the toggle has ON and OFF lettering. It is called a double-pole switch because it handles two hot wires at the same time, rather than one hot wire, as in the single-pole switch. It is commonly used for 240-volt appliances or motors.

Special Switches

If your wall switch makes a loud snap when you turn it off or on, it is an old model. While it may last a good many years more, you might want to replace it with something quieter, more modern, or energy-conserving, such as a dimmer or pilot-light switch. When you are replacing switches, you might want to consider some of the following types.

Dimmer switch. This switch, used to change the intensity of light, has a reostat built into it that reduces the flow of current to the light. Not only does such a switch allow you to create different lighting effects, but it will eventually pay for itself if the light is kept no brighter than necessary. Some models are turned on or off by pushing

Types of Switches

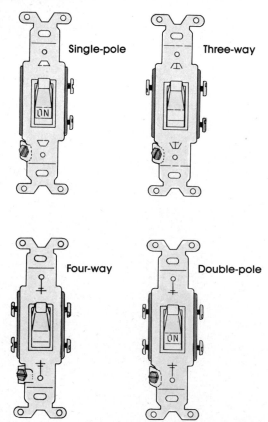

Single-pole

Three-way

Four-way

Double-pole

Time-clock

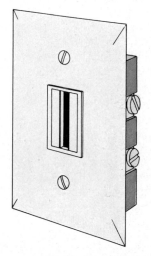

Locking

the dial; others must be turned all the way to OFF. They are primarily used on incandescent lights but are also available for fluorescent lights.

Mercury switch. Mercury conducts electricity. When this switch is turned on, mercury in the switch slides down to make contact with the hot wire. When turned off, the switch slides the mercury away from the contact point. It is absolutely silent and, because it has few moving parts, is commonly guaranteed for 50 years. Mercury switches must be put in place right side up; that end is marked "top." Make sure the switch is vertical for a smooth operation.

Quiet switch. This is the standard switch now sold, and if your house is less than 20 years old, it probably has them. If you still have switches that click loudly each time they are moved, you might want to replace them with this inexpensive model.

Lighted-handle switch. If you have a room where you are constantly groping for the switch in the dark, this is the replacement you need. A miniature neon bulb keeps the toggle glowing in the dark. The bulb uses virtually no electricity.

Pilot-light switch. Many basement or garage lights are controlled from inside the house. The handle or a pilot-light switch glows when the outside light is on, giving you a gentle reminder to turn it off if not in use.

Time-delay switch. If you have an automatic garage-door opener, a light goes on when the door is opened and goes out shortly after it is closed. That may not give you enough time to get groceries out of the car and into the house. The time-delay switch will go off about 45 seconds after you flip it to "delay," giving you time to get the bags inside.

Manual-timer switch. This switch has a spring-wound timer that can be set from a few minutes up to 12 hours. If you have an electric skillet or similar appliance that you want to be turned off while you are out shopping, you can set this timer to the desired time, plug in the skillet, and be on your way.

Time-clock switch. This switch will turn the lights on and off for you at preset times. If you are away from home in the evenings periodically, you can wire it into circuits where you have lamps and a television to make it appear that you are home when you aren't. In addition to the clock mechanism, this switch has a push button for turning lights on and off.

Clock switch. This switch is not only a conversation piece but is handy in several rooms, such as the kitchen, bathroom, or dressing room. A digital clock built into the switch reminds you not to be late.

Locking switch. To keep small children from turning on power tools in your shop, consider using the locking switch. Rather than a handle, it has a key you insert in the switch to turn it on or off. Wire this switch into a circuit that has outlets for your tools, and when you lock it off, no one can turn on a tool.

Time-delay

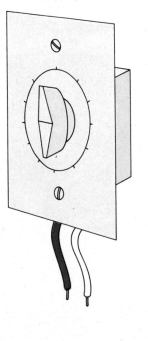

Dimmer

Clock

Pilot-light

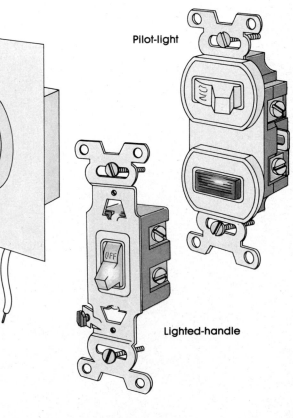

Lighted-handle

Reading a Switch

The type of switch you install depends not only on what you want it to do but also on how it is rated. This important information is stamped on the metal mounting strap, or yoke.

First, look for the marking indicating Underwriters Laboratories' testing and approval for safety standards. A switch without the UND. LAB. marking may be cheaper but may also be of lesser quality.

Next, check the amperage rating. Switches should not be used for amperage higher than that stamped on them. Most lighting circuits use No. 14 wire, which has a maximum ampacity of 15 amps. If that is the case in your house, then use a switch marked "15 amps-120 volts." If you are putting in a new switch where No. 12 wire is used, which has an ampacity of 20 amps, you should still confirm that your local codes permit this.

New switches are stamped with "AC ONLY," which means they can be used only with alternating current, the only type now available in residences. In older houses, the switches that click loudly when moved were designed for either AC or DC (direct current). These switches have no markings to indicate that they can be used for AC or DC. You can replace them with switches marked "AC ONLY."

Finally, but equally important, if you have aluminum wiring in your house, make sure the yoke is stamped with the letters "CO/ALR." This means the switch can be safely used with aluminum wire. If it is marked "CU/AL," it can be used only with copper wire or copper-clad aluminum wire (see page 15 on aluminum wiring).

Reading a Switch

Everything you need to know to choose the proper switch is either stamped into the mounting yoke or molded into the back of the plastic case. Study it all carefully.

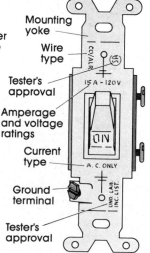

Mounting yoke
Wire type
Tester's approval
Amperage and voltage ratings
Current type
Ground terminal
Tester's approval

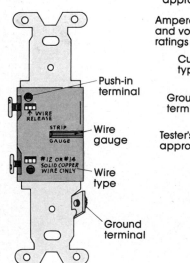

Push-in terminal
Wire gauge
Wire type
Ground terminal

How Switches are Wired

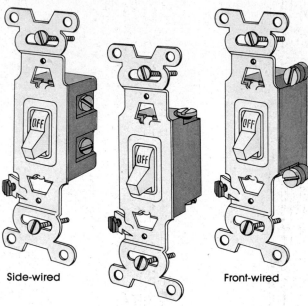

Side-wired

End-wired

Front-wired

Back-wired (rear view)

Combination back-wired and side-wired (rear view)

Attaching Wires to Switches and Receptacles

The three most common types of switches and receptacles are the side-wired, the back-wired, and a combination of both. They all work equally well, the difference being primarily in the manufacturer's design. With the side-wired switch, the cable wires must be wrapped around the screw terminals on the side, as described below. The back-wired switches are somewhat easier to use and a little more expensive. The wires are stripped on the end and pushed into the appropriate slots. With the combination units, you can use either the terminal screws or the push-in terminals. In addition, front-wired and end-wired switches are also available, but they are not commonly used.

Wires are connected to switches or receptacles in two basic manners: under a binding screw terminal or in a push-in terminal. Aluminum wire can be used under a screw terminal only if the device is marked "CO/ALR." **Aluminum wire cannot be used at all in push-in terminals.**

Binding screw terminal. Connecting wire to a binding screw terminal is a relatively simple procedure but it must be done correctly. Loosen the screw so the stripped end of the wire will fit easily under it. Strip away only enough insulation so that the wire will wrap two-thirds to three-quarters of the way around the screw. Use long-nosed pliers to bend the wire in a small loop and hook it around the screw. The loop must face in a clockwise direction so that as the screw is tightened it will pull the loop tight about it. Make sure that the tip of the wire is not bent up or down. Do not bring the wire so far around that it overlaps, which will cause a poor connection. Professionals strip off about 1½ inches of wire, wrap it around the screw, snug down the screw, wiggle the wire back and forth at the three-quarters turn mark to break it there, and then snug down the screw again. Finally, make sure the screw is thoroughly tightened down. A poor connection causes a drop in voltage. Worse, as the electricity tries to jump the gap, it could result in sparks that might cause a fire.

Terminal screws on switches are mounted in different ways, and you may find one type easier to work with than another. Despite this, keep in mind that switches and their mounting straps are all standardized so they will fit any kind of box, and any type of faceplate will fit over them.

Push-in terminal. To connect this kind of terminal, simply strip off about ¾ inch of wire (as much as shown on the strip gauge on the reverse side of the switch or receptacle), and push the wire into the opening. A spring holds the wire in place and makes the contact. If you must remove the wire for any reason, push a screwdriver tip into the slot just above the wire hole to release the spring.

Remember, push-in terminals should not be used with aluminum wire. They are generally acceptable for copper-clad aluminum wire, but check your local codes first.

One screw, two wires. Sometimes you have two wires and only one screw terminal to put them on. A common example is connecting the grounding wire to the switch or receptacle. The incoming and outgoing grounding wires cannot both be put under the same grounding screw because you can't get a proper connection this way, and it is against the Code. The solution is to use a wire nut and a *pigtail*, a six-inch-long piece of wire the same size as the other two and stripped at both ends. Twist the two wires and the pigtail together with a wire nut. Make a loop in the other end of the pigtail and tighten it down under the grounding screw.

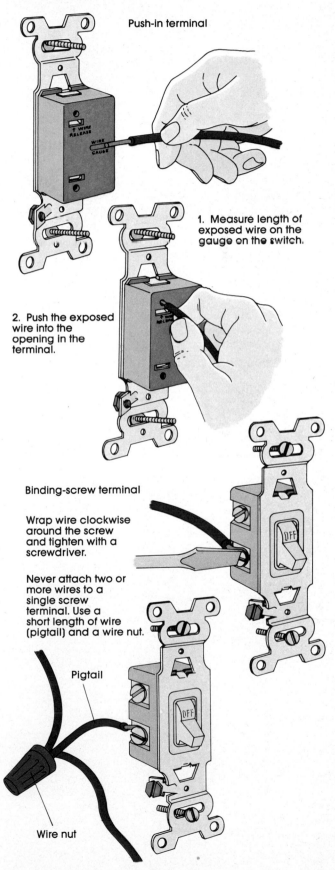

Connecting Wires to Switches

Push-in terminal

1. Measure length of exposed wire on the gauge on the switch.

2. Push the exposed wire into the opening in the terminal.

Binding-screw terminal

Wrap wire clockwise around the screw and tighten with a screwdriver.

Never attach two or more wires to a single screw terminal. Use a short length of wire (pigtail) and a wire nut.

Pigtail

Wire nut

Replacing a Switch

Switches normally last for years and years, but they can wear out. Even if they don't, you may want to replace them with something more modern, particularly if your house still has the loud snap switches. The switches you are most likely to find in your house are single-pole, three-way, and possibly four-way switches. All these switches must be wired in a certain way, as we will discuss here. But before you change a switch because you think it is malfunctioning, check to be sure that it is.

Identifying a malfunction. If a light fails to go on when you flip the switch, the first thing to check is that the bulb is still good. If it is, check that the fuse or breaker on that circuit is still working properly. If both these tests are positive, check out the switch.

First, cut the power on that circuit by removing the fuse or turning the breaker switch to OFF. Next remove the wall plate over the switch. Before touching the switch, use your voltage tester to make sure no power is coming to it. Place one probe on each of the screws in turn while the other probe rests on the metal box. If you have plastic boxes, put the other probe on the bare ground wire. If the switch has push-in terminals, put one probe in each release slot while the other probe is grounded on the metal box or the bare grounding wire. At no time should

the neon light in the voltage tester glow. If it does, the circuit is still hot. Go back and find the proper fuse or breaker for that circuit. If the circuit is dead, unscrew the top and bottom screws on the yoke and pull the switch out of the box. To check that the switch is malfunctioning, use the continuity tester as described on page 12.

Replacing a single-pole switch. When only one switch controls a ceiling light or an outlet with a lamp plugged into it, you have a single-pole switch. To replace it, shut off the power on that circuit, remove the wall plate, and use the voltage tester to make sure the circuit is dead (see page 11). Remove the top and bottom screws from the yoke and pull the switch out of the box.

Note that only two black wires and a bare grounding wire are attached to the old switch. The two white wires in the box are joined by a wire nut. Do not bother the white wires. Loosen the screws that hold the black wires. If the terminal is a push-in type, push a small screwdriver into the slot above the push-in terminals to free the wire.

Put the black wires back under the screws on the new switch and tighten down. Make the loop three quarters of the way around the screw and check that no wire is sticking out from under it. With the single-pole switch, it does not matter which black wire goes under which

Replacing a Single-Pole Switch

Old switch

New switch

Grounding wire

Switch loop

Old switch

New switch

Grounding wire

screw. If it is a push-in switch, just make sure the wires do not go in the holes labeled "white."

If you are changing the switch from one with screw terminals to a push-in type, clip off the bent ends of the wire and then use the strip gauge on the back of the switch to strip off the correct amount of insulation before inserting the wires in the terminals.

The bare grounding wire should be connected to the green hexagonal screw on the new switch. If no screw exists on the replacement switch, use a pigtail (see page 39) to connect the switch to a metal box.

If you pull the switch from the box and see just two incoming wires attached to the switch and no wires continuing on, it is wired for a *switch loop* (see page 44). The white wire should go on the bottom terminal screw, and the black wire should go on the top one. In this particular case, the white wire is also hot and should be so identified by a dab of black paint or a wrapping of black electrician's tape. If it is not already marked, do so when you put in the new switch. Power comes in on the white cable and, when the switch is thrown to ON, continues on through the now hot black wire to the receptacle or light fixture.

When you install any new switch, make sure that it is vertical. The box may be crooked, but the mounting screw slots are wide enough so you can adjust the switch if necessary. Note that each end of the yoke has a pair of "ears." If the box is recessed, these ears will keep the switch flush with the wall. If the box is already flush with the wall and the ears are in the way, snap them off with a pair of pliers.

Replacing a three-way switch. Three-way switches are always used in pairs, normally to control a light from either end of a room or hallway, or at the top and bottom of stairs. If a light doesn't work, and checking the bulb plus the fuse box or breaker switch indicated the problem might be the switch, you must first determine which switch it is. Shut off power to that circuit, remove one switch, and check it with a continuity tester, as described on page 12.

Once you have located the defective switch, remove it from the box. Use the voltage tester to ensure that all power to that switch is shut off before working on it (see page 11). Before removing any wires, put a piece of tape on the wire attached to the common terminal. Either it is marked "common," or the screw is a darker color than the other two. The common terminal screw is either black or copper-colored to distinguish it from the other two silver- or brass-colored screws.

Remove the other wires from the switch. These will include a hot black wire, a hot red wire (traveler wire), a white neutral wire, and a green or bare grounding wire. Reconnect these wires in the same fashion to the new three-way switch and then connect the marked wire to the common terminal.

If the bare grounding wire was not attached to the green hexagonal screw on the old switch, use a pigtail (see page 39) and a wire nut to connect the switch to the grounding wires and another pigtail to a screw in the metal box (see diagram).

Replacing a Three-Way Switch

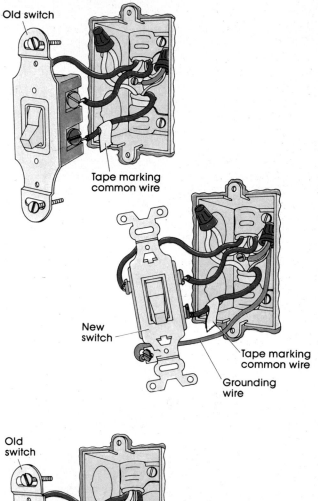

Old switch

Tape marking common wire

New switch

Tape marking common wire

Grounding wire

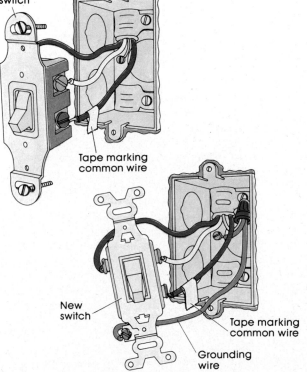

Old switch

Tape marking common wire

New switch

Tape marking common wire

Grounding wire

Replacing a four-way switch. A four-way switch is used between two three-way switches to provide additional points in a room from where the lights can be turned off. In some cases more than one four-way switch may be installed. Remove the wall plates on all the switches along that circuit to identify the four-way switches. They have four terminal screws, while the three-way switches have only three.

Once you have identified the faulty four-way switch (see page 12 on using continuity tester), make sure power is off on that circuit and remove the switch from the box. Note that all four wires connected to the switch are hot. Connected to the top terminal screws will be a red (possibly black) and a white wire marked as black (or possibly a black wire.) The same will be true on the bottom terminal screws. The easiest way to replace the switch is to remove the wires to the top terminals and put them on the top terminals of the new switch. It does not matter which wire goes to which screw, as long as both are on top. Now do the same with the wires on the bottom terminal screws. Do not disturb the black wires joined by a wire nut in the box. Attach the grounding wire to the switch's green hexagonal screw and to the metal box with a pigtail, as shown.

Put the switch back in the box, put the wall plate back on, turn the power on, and test your work. If it doesn't work, that switch may have a different wiring system. In that case, shut off the power again and remove the switch from the box. Loosen the two screws on just one side of the box, reverse the wires, and test the switch again.

Installing a Dimmer Switch

Dimmer switches either come with a dial to turn the lights up or down, or they appear to be a standard wall switch with the amount of light controlled by the relative movement of the switch. These switches can handle only a certain amount of wattage, which is stamped on the switch. Dimmers with knobs usually can handle up to 600 watts, while those with toggle switches have a 300-watt capacity. Either kind usually has more than enough, but do not exceed the limit.

Dimmer switches are normally used to control overhead incandescent lights. Standard dimmers cannot be used on fluorescent lights, but special ones are available. Simple dimmers that are easily attached in the middle of a lamp cord to control floor or table lamps are also available in most hardware or electrical supply stores. Wall-mounted dimmer switches are available in single-pole and three-way switches. When used in a three-way switch, however, only one of the two switches will dim the light. Both continue to turn the lights on or off.

A single-pole dimmer switch is hooked up just like a standard single-pole switch (see page 40). Some dimmers come with switch leads (wires) already attached to the mechanism. These are joined to the existing wires (black to black, white to white) with wire nuts.

When installing a dimmer on a three-way switch circuit, first mark the black common wire with a piece of tape. Attach it to the black lead from the switch with a wire nut. Then attach the other two leads with wire nuts,

Replacing a Four-Way Switch

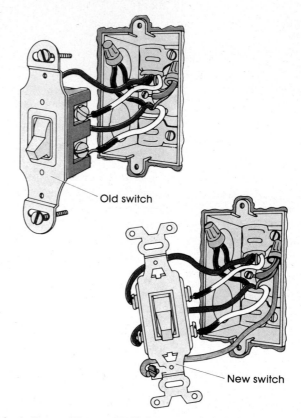

Old switch

New switch

Installing a Dimmer Switch

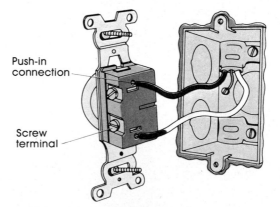

Push-in connection

Screw terminal

Installing a Three-Way Dimmer

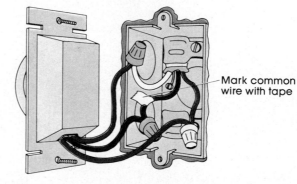

Mark common wire with tape

Installing a Pilot-Light Switch

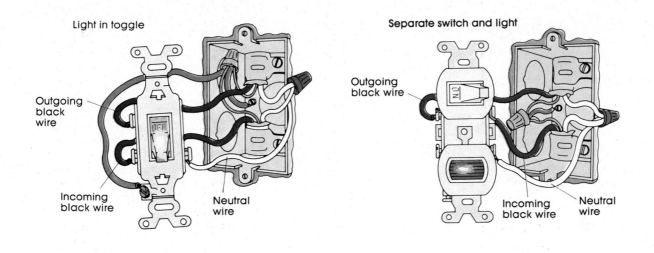

Light in toggle

Outgoing black wire

Incoming black wire

Neutral wire

Separate switch and light

Outgoing black wire

Incoming black wire

Neutral wire

one to the white grounded wire and one to the red traveler wire.

Installing a Pilot-Light Switch

The pilot-light switch, with its small light that advises you when the light outside or in another room is on, must be installed in middle-of-the-run wiring (see page 47). Installation is similar to that of a single-pole switch, but you must tap into the white neutral wire for the pilot light. Pilot-light switches are made in two basic styles: with a light in the toggle or with a separate light next to the toggle. Each is wired slightly differently.

If the light is in the toggle, it will have one silver-colored and two brass-colored terminal screws. Connect the incoming black wire to one brass-colored screw and the outgoing black wire to the other. Attach one end of a white jumper wire to the silver-colored screw, and connect the other end to the white neutral wire with a wire nut. If the light stays on when the switch is off, reverse the two black wires.

With a separate light switch, there will be one silver-colored and three brass-colored terminal screws. You will see that on one side two brass-colored terminal screws are joined by a metal strip. Attach the outgoing black wire to one of these two screws. Attach the incoming black wire to the brass-colored screw on the other side. Attach the white jumper wire to the silver-colored screw, and then tie it with a wire nut to the white wires in the box.

Installing a Time-Clock Switch

Like pilot-light switches, time-clock switches must be installed in middle-of-the-run wiring. After making sure the power is off, remove the old switch. Turn the power back on and use a voltage tester to find the incoming hot black wire. Turn the power back off and mark the wire with tape. Remove the wire nut from the white neutral wires.

Attach the special mounting plate that comes with the timer switch to the box. Next, using wire nuts, join the incoming black wire to the black wire on the timer switch. Join the red wire on the switch to the outgoing black wire. Finally, join the two wires in the box to the white jumper wire from the switch. Put the switch in place, turn the power on, and test the switch according to the manufacturer's instructions.

Installing a Time-Clock Switch

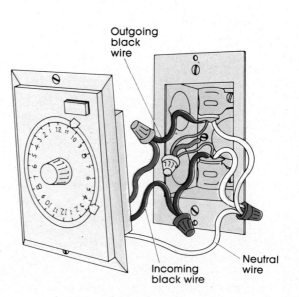

Outgoing black wire

Incoming black wire

Neutral wire

RECEPTACLES

A Switch Loop

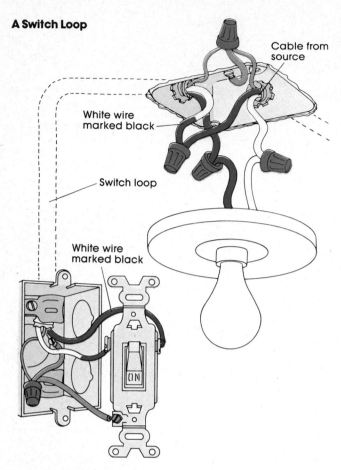

Cable from source

White wire marked black

Switch loop

White wire marked black

Installing a Switch Loop

A switch loop is used when the incoming power passes through the light fixture before arriving at the switch that controls the fixture. Although this is the reverse of standard wiring, the switch loop is a common technique used when adding on to an existing circuit. For example, it can be used to put an overhead light in a new bedroom addition.

So far you have learned that the black wire is always hot and the white wire is always neutral. In this case, the white wire will become hot, and because of this it must be so indicated by wrapping each end near the terminal screws with black electrician's tape, or by painting the ends black.

In the switch loop, an incoming hot black wire and white neutral wire arrive at the light fixture before they reach the switch. A black wire and a white wire also come from the switch to the light fixture. If the incoming hot black wire were attached to the light fixture, the light would always be on. To prevent this, the hot wire must first loop through the switch. To do this, connect the incoming hot wire to the white wire from the switch with a wire nut. The white wire becomes hot (this is the one you must mark with black electrician's tape). The incoming white neutral wire is attached to the light fixture and marked. Finally, attach the black wire from the switch to the second screw on the light fixture. Incoming electricity now goes first to the switch then back to the fixture.

If a fuse blows every time you plug an appliance or lamp into an outlet, there is probably a short circuit. It may be caused by loose wires, or the receptacle may be faulty.

To check, shut off power on that circuit, remove the wall plate, and pull the receptacle out of the box. Use a voltage tester to make sure power is shut off (see page 11). If a terminal screw appears loose, tighten it and the others, and reinstall the outlet. If the fuse blows again when you plug in a lamp, replace the outlet.

Three-hole grounding-type outlets are now required in all new house construction and are advised for any rewiring work. If the outlet you are replacing is the older, two-hole, ungrounded variety, consider replacing it with the three-hole receptacle. To do this you must attach a grounding wire to the outlet and to the switch box if it is metal. If the box is plastic, it will not conduct electricity and need not be grounded. If your house is old and was poorly wired to begin with, it may have only a hot and neutral wire coming to the switch, and no grounding wire. In such a case, do not install a grounding-type plug. There are two basic types of receptacles: side-wired and back-wired.

The *side-wired* receptacle has four terminal screws: two brass-colored and two silver-colored. The hot wires, black or red, go on the brass-colored screws; the white neutral wires go on the silver-colored screws. The green screw at the bottom is for the grounding wire.

The *back-wired* or push-in receptacle has holes in which to insert the wires. A strip gauge on the back of the outlet shows you how much insulation to strip off the end of the wire. The word "white" is stamped on one side of the plug, indicating that the white neutral wires go in there. The hot red or black wires go in the other side. To release the wires, push a screwdriver into the slot above or below the holes. Some receptacles have both push-in holes and terminal screws. Most receptacles are duplex receptacles, so called because they have two units. Both units are hot at the same time because they are joined by a brass tab on one side that passes the current from one side to the other. On the other side is a chrome tab that keeps current flowing in the white neutral wire. If

Types of Wiring on Receptacles

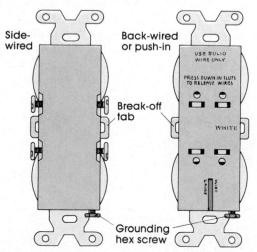

Side-wired

Back-wired or push-in

USE SOLID WIRE ONLY

PRESS DOWN IN SLOTS TO RELEASE WIRES

WHITE

Break-off tab

Grounding hex screw

Reading a Receptacle

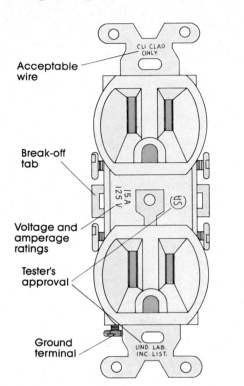

- Acceptable wire
- CU CLAD ONLY
- Break-off tab
- 15A 125V
- Voltage and amperage ratings
- Tester's approval
- Ground terminal
- UND. LAB. INC. LIST.

Replacing a Receptacle

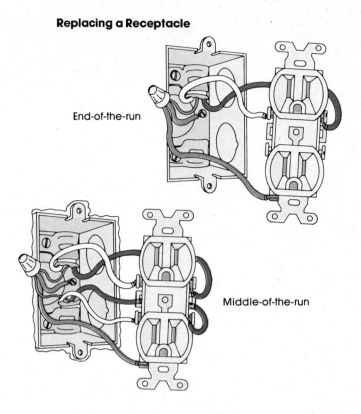

End-of-the-run

Middle-of-the-run

the brass tab is broken off by twisting it with a pair of pliers, only that outlet which is actually wired will be hot. The current will not pass to the other. This is done for *split-receptacle* wiring with three-wire cable, a complex process not covered in this book (a simpler alternative for a three-wire circuit is given on page 63).

Finally, note that all outlets currently being sold are designed to handle three-prong plugs. The U-shaped hole is the grounding connection. All new installations must use these. But if you have an older house, it may have only two-slot receptacles. If that is the case, it is a good idea to replace them with three-slot grounding outlets, but *only* if the existing receptacles are already grounded (check by turning off the power, removing the faceplate, and looking for a third, bare copper wire, which is the grounding wire). If your present receptacles are not grounded, it is against the Code to install three-slot grounding receptacles. This would give the erroneous impression that your wiring system was grounded.

Reading a Receptacle

A receptacle should be stamped with the UND. LAB. marking, which indicates that the outlet has passed Underwriters' rigid safety standards. The receptacle will also carry a stamp indicating the maximum amperes it can handle and the number of volts. Typically, it reads "15A-125V." On the outlet or the mounting yoke will be stamped a series of letters indicating what type of wire may be used with this receptacle. If it reads "CU" or "CU CLAD," use only copper or copper-clad aluminum wire. If your house has aluminum wiring, you must use a receptacle stamped "CO/ALR."

Replacing a Receptacle

End-of-the-run receptacle. If this is the last receptacle on the circuit, only two wires will be coming into the box. The hot black wire is attached to one brass-colored terminal screw, and the white neutral wire is attached to the silver-colored screw. The bare grounding wire is attached to both the green terminal screw on the receptacle and to the metal switch box by the use of pigtails and a wire nut, as shown.

If armored cable is used, there will be no bare grounding wire. In this case, ground the receptacle by running a jumper wire from the green grounding screw to a machine screw in the back of the metal box.

Middle-of-the-run receptacle. In this case, there are four wires in the box: a hot black and a white neutral wire coming in, and the same going out to the next receptacle. In addition, there are incoming and outgoing grounding wires.

To wire this receptacle, attach the black wires to the two brass-colored screws and the white wires to the silver-colored screws. Attach one green pigtail wire to the green terminal screw on the receptacle. Attach another one to the metal switch box with a machine screw. Join these two and the incoming and outgoing grounding wires with one wire nut.

Middle-of-the-run receptacle with armored cable. The armored cable serves as the grounding wire in this box. To wire this receptacle, attach the black wires to the brass-colored screws and white wires to the silver-colored screws. Using a green pigtail wire, attach one end to the green terminal screw on the receptacle and the other end to the metal box with a machine screw.

Installing a Switch-Receptacle Combination

If you need a receptacle where there is now only a switch, you can use this combination. But it must be middle-of-the run wiring because an outlet always must be hooked to a white neutral wire to complete the circuit. This combination has one silver-colored, one copper, and two brass-colored screws. To start, attach a jumper wire to both neutral wires with a wire nut and to the silver-colored screw.

To have the outlet and the light controlled by the switch, attach the black power wire to the copper screw and the black switch wire to either brass-colored screw.

If you want the outlet hot all the time, attach the black power wire to either brass-colored screw and the black switch wire to the copper screw.

Connect a green pigtail grounding wire to the back of the metal switch box and another one to the green hexagonal screw on the switch receptacle. Tie the two pigtails into the bare grounding wires with a wire nut.

Installing a Light-Receptacle Combination

Some older houses have a shortage of outlets. But a light over the stove or sink or some other place controlled by an existing switch can become both a light and a receptacle. This kind of light is sold with the fixture wires already installed. You can install this receptacle in end-of-the-run wiring as follows:

At an end-of-the-run connection, the outlet will be

Switch-Receptacle Combination

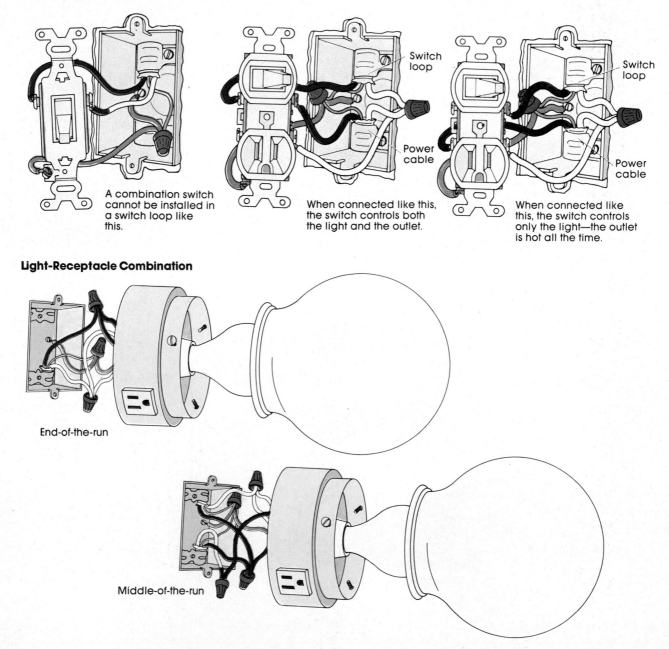

A combination switch cannot be installed in a switch loop like this.

Switch loop

Power cable

When connected like this, the switch controls both the light and the outlet.

Switch loop

Power cable

When connected like this, the switch controls only the light—the outlet is hot all the time.

Light-Receptacle Combination

End-of-the-run

Middle-of-the-run

hot only when the light is on. The light and the receptacle need both hot and neutral wires. Using a wire nut, attach the two black wires from the fixture to the incoming hot black wire. Attach the two white neutral wires from the fixture to the incoming white neutral wire with a wire nut. Connect a green grounding jumper wire to the back of the metal switch box and then, with a wire nut, link that to the incoming bare grounding wire and the green grounding wire from the fixture.

At a middle-of-the-run connection, the outlet will be hot all the time. Using a wire nut, connect the black wire from the receptacle part of the fixture to the two black cable wires. Connect the white wire marked black to the black wire from the light. Attach the two neutral wires from the fixture to the incoming white wire. Connect a green grounding jumper wire to the back of the switch box and join it with a wire nut to the incoming bare grounding wire and the green fixture wire.

Types of Plugs and Receptacle Slots

Ungrounded Two-Prong Plug (120 volts)
This plug is commonly used on lamps, toasters, irons, and other small appliances. In many instances, one prong is slightly larger than the others; the smaller prong is hot and the larger one is neutral. The plug can be inserted into an outlet only one way. If the receptacle is wired properly, the incoming hot black wire is connected to the side of the outlet with the shorter and narrower slot. The white neutral wire is connected to the longer and wider slot.

Grounded Three-Prong Plug (120 volts)
The three-prong plug indicates that the cord contains a grounding wire. When it is plugged into a properly wired three-slot outlet, any short circuit that might occur in the tool or appliance will be directed through the grounding wire to the grounding wire in the receptacle, rather than through you. Three-prong plugs are usually for 15 amps. (The 20-amp grounded plug has one prong angled differently from the others and will fit only a 20-amp three-slot outlet, as shown.)

Grounded Plug (120/240 volts—30 amps)
This plug and receptacle is designed specifically for clothes dryers. Note that the L-shaped grounding plug will only fit into a similarly shaped slot in the outlet. This outlet is wired with both 120 and 240 volts. The 240 volts are for the dryer's heating coils, and the 120 volts are for the inside light and the timing mechanism.

Grounded Plug (120/240 volts—50 amps)
This plug and outlet is used primarily for electric ranges. The prongs are all angled a specific way and will fit only a matching outlet. The 120 volts are used to power the clock, lights, and burners when on low settings. The 240 volts power the oven and the burners when on higher settings.

Grounded plug (240 volts—30 amps)
This type of plug, again with a set of prongs that will fit only a matching outlet, is commonly used for air conditioners and water heaters.

Types of Plugs and Outlets

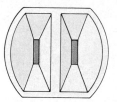

Ungrounded two-prong (120 volts)

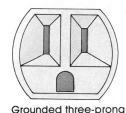

Grounded three-prong (120 volts—15 amps)

Grounded three-prong (120 volts—20 amps)

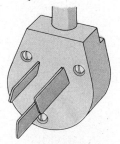

Grounded three-prong (120/240 volts—30 amps)

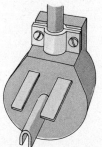

Grounded three-prong (120/240 volts—50 amps)

Grounded three-prong (240 volts—30 amps)

SOME COMMON REPAIRS

A small repair job can prevent a big one later on. Try your hand at replacing a plug, rewiring a lamp, installing a ceiling light, or fixing a fluorescent lamp or a doorbell.

If you own your own house, you know that endless repairs and maintenance are required. You may have put off small electrical problems because they seem so complicated. However, some of those small problems are dangerous if left unattended, such as frayed cords, faulty plugs, and lamps that don't work. If you take a careful look at all the lamps and cords in the house, you are likely to find some that need repairing or replacing. Now is probably as good a time as any to fix them. And this is a good way to gain experience.

Cords

There are several different types of cords for lamps, small appliances, heaters, or power tools. They are all lumped under the general term of *flexible cords.* The wire is generally 16- to 18-gauge and stranded rather than solid, which makes it more flexible. Most of these cords contain only two wires but the cords on power tools may have a third, which is the grounding wire. There are usually only one or two layers of light insulation, which helps keep the cord flexible. But this also means the insulation is more subject to wearing through and exposing the wires, which can give a shock or cause a fire. A temporary solution to worn insulation is to wrap the cord thoroughly with plastic electrician's tape. But a better solution is either to remove that worn part and reinstall the plug, or to replace the entire cord.

Fixture cord. This is commonly called "zip" or type SPT. It consists of two wires in a light thermoplastic insulation.
Heater cord. There are two basic types: HPN, which looks like a heavy-duty zip cord, and HPD, which has a woven cotton exterior covering over a layer of asbestos filling around the wires.
Vacuum cleaner cord. This type, SVT, has a rubber in-

sulating skin and a paper interior layer. The two wires are encased in a light thermoplastic covering.
Power cord. This type of cord, SJT, has a thermoplastic outer layer. The two or three wires inside (depending on whether it has a grounding plug or not) are individually encased in a light thermoplastic layer.

If you look carefully at all the cords in your house, garage, and shop, you may spot some problems that need immediate attention. Check for any cord that is frayed, melted, cracked, or broken, exposing the wires. You may also find a cord coming loose from its plug.

If the problem area is near the plug or appliance, unplug the cord from the receptacle, cut the cord behind the damaged area, and then reattach it to the plug or appliance. If the damaged area is in the middle of the cord, you should replace the entire cord. Take the old one to a hardware or electrical supply store to buy a cord of the same type and length. The same length is important; a longer cord would mean more resistance and less efficiency in the electrical appliance.

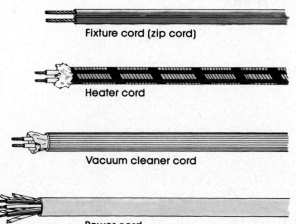

Fixture cord (zip cord)

Heater cord

Vacuum cleaner cord

Power cord

◄

Simply replacing a damaged plug may be all it takes to make an appliance useful again.

REPLACING PLUGS

If plugs become cracked or the prongs break, they must be replaced. A plug that feels warm when in use is faulty and should be replaced. If the plug just doesn't fit tightly in the outlet, bend the prongs a little farther apart before deciding to replace it. If you must replace it, note what type it is and buy a matching one. You probably have most of the following kinds of plugs somewhere in your home (for additional types, see page 47).

Screw terminal plug. This is one of the most common types of plugs in the house. To replace it, first pry off the cardboard insulator at the base of the prongs. Loosen the two screws holding the wires. Don't try to completely remove the screws since they stop automatically at a certain point. Remove the wires from under the screws, untie the knot, and remove the old plug. If the cord has been damaged near the plug, cut the cord behind the damaged section and then part the wires about two inches. If there is an outer insulation layer, remove that first by cutting around it with a knife. Slip the wires through the new plug and then remove ½ to ¾ inch of insulation from the end of each wire. Tie the Underwriters' knot, as shown. Twist the stranded wire on each end clockwise and then wrap each end clockwise under a screw and tighten it down. Unless the plug is a polarized type, it doesn't normally matter which wire goes to which screw.

Polarized plug. This type is replaced just like the standard screw terminal plug with one exception: the black hot wire must be attached to the narrower prong. One prong is always narrower on a polarized plug. Some appliances, such as televisions, are wired in such a way that the prongs must match the incoming power. The narrower prong is the hot one and the wider prong is neutral. These will fit into a receptacle only one way. The receptacle also must be wired accordingly, with the hot wire connected to the smaller slot.

Three-prong plug. This plug is wired in a manner similar to the standard screw terminal plug, but the black hot wire must be attached to the brass-colored terminal screw and the white wire to the silver-colored terminal screw. The third wire, the bare grounding wire, is attached to the green grounding screw terminal. In many cases, particularly on power tools, the plug is encased in the same insulation as the cord. If this type of plug is damaged, cut the cord just behind the plug, throw the old one away, and then replace it with a standard three-prong screw terminal plug.

Appliance plug. Remove the center screw that holds the two halves of the plug together. Loosen the two terminal screws and remove the old plug. Wind the stripped end of the wire clockwise around the terminal screws on the new plug and tighten. Keep the spring in place in the new plug and then put both halves together and tighten the center holding screw.

Self-connecting plug. This type is used only with zip cord. Do not separate the wires but make sure the ends are cut square. Pull out the prongs on the plug and push the wire into the unit as far as it will go. Squeeze the prongs together firmly and slip the cover over the unit.

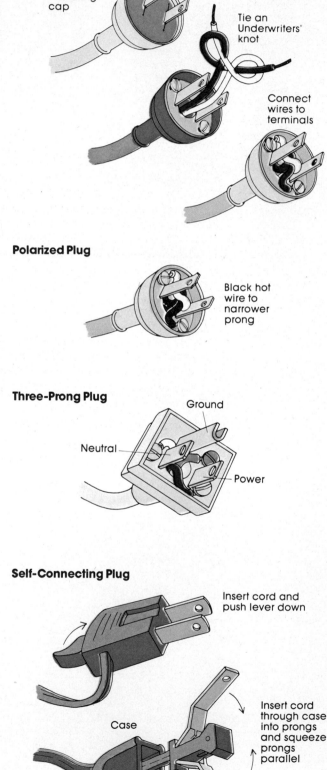

Screw Terminal Plug

Remove insulating cap

Tie an Underwriters' knot

Connect wires to terminals

Polarized Plug

Black hot wire to narrower prong

Three-Prong Plug

Ground

Neutral

Power

Self-Connecting Plug

Insert cord and push lever down

Case

Insert cord through case into prongs and squeeze prongs parallel

INCANDESCENT FIXTURES

Rewiring an Incandescent Lamp

Lamps with incandescent bulbs are all quite similar, whether they are large floor lamps or small table models. A lamp has only a few basic parts that can cause a malfunction: the bulb, the socket, the switch, the cord, and the plug. Understanding how a lamp is put together not only allows you to make any repairs necessary, but also to make your own lamps from old bottles, pieces of driftwood, or anything that strikes your fancy. All the parts to put a lamp together from scratch can be found in kit form in hardware and electrical stores.

Taking Apart a Lamp

1. Unplug the lamp and remove the bulb.
2. Remove the shade. It may be necessary to unscrew the finial at the top of the harp to free the shade. If your lamp has a two-piece harp, slide the two metal sleeves up at the base of the harp, squeeze the harp at the base, and lift it out.
3. The socket consists of four pieces: the outer shell, the insulating sleeve, the socket with switch and terminal screws, and the socket cap. To remove the shell, press in with your thumb where the shell is stamped "press" and lift up. If the shell will not come off, pry it up with a screwdriver. Remove the cardboard insulating sleeve. Loosen the two terminal screws on the socket and remove the wires. Lift the socket out of the cap.
4. To remove the cap, first loosen the set screw at the base of the cap. Unscrew the cap from the threaded nipple. If there is no set screw and the cap does not readily unscrew, give it a forceful turn.
5. If you must remove the cord from the lamp pipe, tape the new cord to the old and pull it through as you remove the old cord.
6. Part the new cord about two inches back and tie an Underwriters' knot at the top, as illustrated.
7. Strip off ½ to ¾ inch of insulation at the end of the wire, twist the wire clockwise, and wrap it clockwise around the terminal screws. Tighten the screws. Reassemble the lamp.

Lamp Components

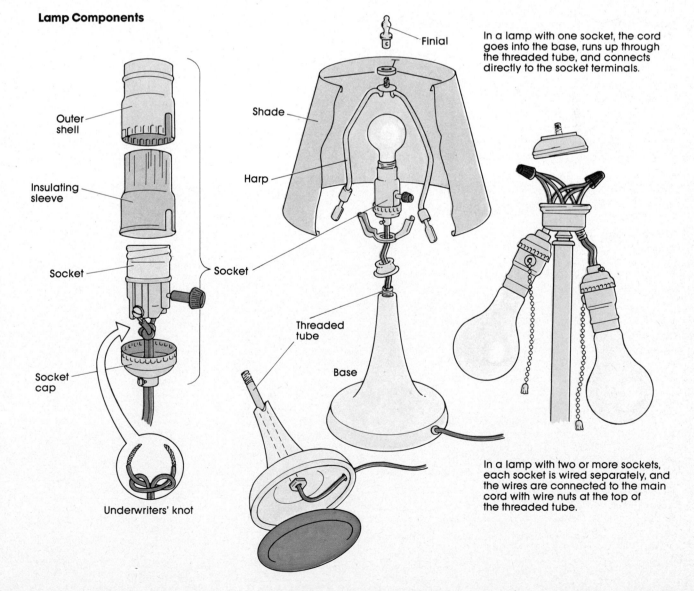

Outer shell

Insulating sleeve

Socket

Socket cap

Underwriters' knot

Shade

Harp

Socket

Threaded tube

Base

Finial

In a lamp with one socket, the cord goes into the base, runs up through the threaded tube, and connects directly to the socket terminals.

In a lamp with two or more sockets, each socket is wired separately, and the wires are connected to the main cord with wire nuts at the top of the threaded tube.

Testing the Socket and Switch

The most common source of lamp failure is the switch, which is the only moving part in the lamp. The socket, however, can also be a problem. Both the switch and the socket can be quickly checked with a continuity tester (see page 12).

To check the socket, first unplug the lamp and remove the bulb. Put the alligator clip on the metal socket and then touch the probe to the silver-colored terminal screw. The tester should light. If it doesn't, replace the socket.

To check the switch, unplug the lamp and remove the bulb. Attach the alligator clip to the brass-colored terminal screw and touch the probe to the metal spring tab inside the base of the socket. The tester should glow when the switch is on and not glow when the switch is off. If the tester shows that the switch is working but the lamp still will not work, you should first make sure the spring tab is bent up enough to make firm contact with the light bulb.

To test a three-position light switch, attach the alligator clip to the brass-colored terminal. Turn the switch to the first ON position and touch the probe to the small vertical tab at the base of the socket. The tester should light. Repeat this process for the second and third ON positions. At the third position, the tester should light on either tab. If any position fails, replace the socket and switch. If they all work and the lamp does not work, double check the cord and plug.

Troubleshooting a Lamp

If a lamp flickers or does not go on at all, run through this quick checklist:

1. Remove the bulb and see if it lights in another lamp you know is working.

2. Check the plug for any cracks or loose wires. Make sure the prongs fit tightly in the receptacle. Spread them slightly for a tight fit. If the plug is cracked, replace it as described on page 50.

3. Check that the cord is not frayed to the extent that bare wires are exposed, which could cause a short circuit. If they are exposed, install a plug on the cord as described on page 49. Make sure the cord wires have not become loose from the plug due to pulling on the wire rather than the plug when removing it from an outlet. Tighten down under the plug's terminal screws if necessary. Self-connecting plugs are quite prone to losing contact with the cord wires.

4. Make sure the spring terminal at the base of the socket is bent up enough to make firm contact with the bulb. If you are in doubt, pry it up slightly with your finger or a screwdriver.

5. As a last resort, disassemble the lamp to inspect the socket and switch.

> **Reminder:** Always unplug a lamp or appliance before you work on it. Remember to pull on the plug itself, rather than the cord.

Installing a New Ceiling Light

One relatively painless way to give rooms a fresh look is to install new ceiling lights. Any room can benefit from this, but good places to start are the living room, dining room, kitchen, and bathrooms. Before you dismantle the existing light, turn off the circuit breaker or remove the fuse for that light. Just turning off the light switch is not sufficient protection.

Discovering how to remove the decorative cover can sometimes be baffling. There may be several screws around the edge or a single, disguised cap nut in the center. If none of these exists, turn the cover plate to the

Hooking Up Ceiling Fixtures

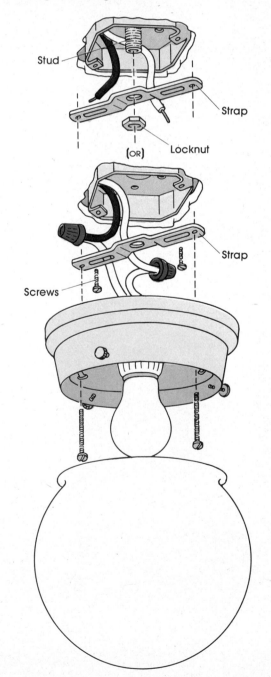

left to unscrew it. Once you have removed the fixture, don't let it hang from the wires. Support it with a piece of bent coat hanger. Then you can use both hands to remove the wire nuts connecting the wires and remove the fixture.

There are three basic styles of ceiling fixture supports, which are illustrated here. Although different, all can be adapted to handle whatever support system your new fixture has. Fixtures are supported from a central threaded stud, screwed to a mounting strap, or both. If there is a central stud in the ceiling box that does not fit the threaded nipple (short piece of threaded pipe) on

your new fixture, ask for a *reducing nut* at the lamp store. The nut has two different-sized threaded ends to fit both the stud and the nipple.

If the new fixture screws to the ceiling box but the screw holes do not line up, get a mounting strap sized to fit inside the cover plate on the fixture. Screw the strap to the ceiling box and then the fixture to the strap. An alternative method is to screw the strap to the ceiling box and then connect the fixture to the strap with a threaded nipple through the hole in the center of the strap. If the fixture wires must be run through the nipple to the fixture, use a hickey, as illustrated.

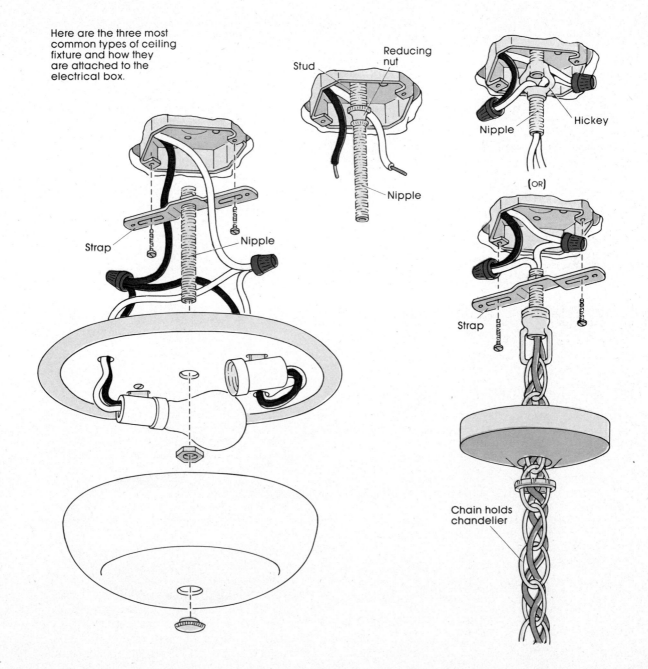

Here are the three most common types of ceiling fixture and how they are attached to the electrical box.

Strap

Nipple

Stud

Reducing nut

Nipple

Nipple

Hickey

(OR)

Strap

Chain holds chandelier

FIXING A FLUORESCENT LAMP

Fluorescent Fixture Wiring

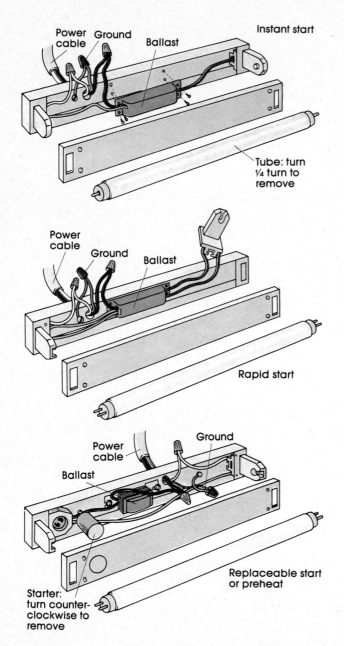

Instant start

Power cable — Ground — Ballast

Tube: turn ¼ turn to remove

Power cable — Ground — Ballast

Rapid start

Power cable — Ground — Ballast

Starter: turn counter-clockwise to remove

Replaceable start or preheat

If you are energy-conscious, begin converting many of the incandescent lamps in your house to fluorescent. They burn brighter and longer and use less electricity. They are particularly practical in kitchens and bathrooms where long tubes can be hidden behind dropped ceilings to provide bright, even lighting.

Efficient as they are, fluorescent lights can go haywire. But if you understand how all the components work, fixing a fluorescent lamp is not much more difficult than repairing a lamp with an incandescent bulb.

Unlike an incandescent bulb, which is lighted by resistance in the filament in the bulb, a fluorescent bulb is lighted by a charge of electricity passing through gas contained in the tube. Standard 120-volt electricity arrives at the fluorescent lamp, then passes through a bal-

last, or transformer, before entering the tube. On two types of fluorescent lamps, a starter is used to help the ballast send a charge of electricity through a tube when the lamp is turned on.

There are three types of fluorescent lamps: *preheat*, *rapid start*, and *instant start*. In the preheat model, the starter is separate from the ballast (see illustration). The starter is built into the ballast on the rapid start. There is no starter on the instant-start model, which is distinguished by its single-pin fluorescent tube.

Changing Tubes

To remove a double-pin fluorescent tube, twist it a quarter turn either to the right or left and gently pull it out. To install a new tube, slip the pins into the slots and give the tube a quarter turn in either direction. On a single-pin model, push the tube back against the spring-loaded socket at one end until the pin at the other end is free, and then lift the tube out. To put in a new tube, put the pin of one end in the spring-loaded socket and push back until you can slip the pin on the other end into the socket.

Replacing the Starter

Starters, which are usually small aluminum cylinders, are either built into the ballast or wired separately. If the starter is in the ballast, remove it by turning it counter-clockwise a quarter turn, then pulling it out. If the starter is separate from the ballast, shut off power to the lamp or unplug it, and then remove the wire nuts connecting the starter to the socket. In both cases, you reverse the process to install a new starter.

Replacing the Ballast

If you have a table-model fluorescent lamp, the ballast is in the lamp base. On a wall or ceiling-mounted lamp, the ballast is a square boxlike device behind the cover plate. To remove it, unplug the lamp, then remove the fluorescent tubes and the cover plate. Cut the two wires running into the ballast and then remove the mounting screws to free the ballast. The new ballast has several inches of wires coming from it. Strip the ends of those wires and the wires from the source and connect the wire nuts. Connect each wire according to the color—blue to blue, red to red, and so forth. A wiring diagram is printed on the ballast to assist you.

Replacing Sockets

Occasionally, the sockets on a fluorescent lamp will be damaged or possibly corroded beyond repair. To replace them, shut off power to the lamp and then remove the wires from each socket. If the wires are held by terminal screws, simply unscrew them; with push-in terminals, the wires can be freed by pushing a small screwdriver into the slot next to the wires. On some models, the sockets are held in place by a mounting screw; on other models they are slipped into position over the ends of the bracket. In the latter case, remove the end cover plate so you can slide the socket out. Take the old sockets with you to the hardware or electrical store so you can match the replacements exactly.

FIXING A DOORBELL

A doorbell is a relatively simple mechanism but it can sometimes stubbornly refuse to ring. There are four main areas to check when a doorbell won't work properly: the bell, the button, the wires, and the transformer.

Doorbells operate at very low voltage, somewhere between 6 and 24 volts. Power arrives in 120 volts at the transformer, where it is stepped down to low voltage. Because 120-volt power arrives at the transformer, when you work with it, shut off all power on that circuit. Otherwise, in checking out a faulty doorbell, you usually need the power on. The voltage is so low you won't feel anything in touching the wires.

If the bell does not ring, first make sure that the fuse or circuit breaker is functioning properly. If that is not the problem, check each of the four problem areas:

1. Have someone push the doorbell while you listen closely. If you can hear a buzz or faint sound, remove the bell and make sure that the clapper is not gummed up. Clean the clapper and bell thoroughly.

2. Remove the cover from the button, loosen the wires from the terminal screws, and cross the wires. If the bell rings, the problem is in the button. Clean the contacts by scraping them down to bare metal or replace the button.

3. Check all the wires. They are normally No. 16 or No. 18, a lightweight wire that can easily be broken or frayed. Wrap any frayed areas with electrician's tape and repair any breaks by stripping the ends and connecting them with wire nuts.

4. Check that the terminal screws at the button and the transformer are tight. If they are greasy or dirty, or have been painted over, remove the wires and clean the terminals with sandpaper.

These basic steps are normally enough to get a doorbell ringing again.

Doorbell Wiring

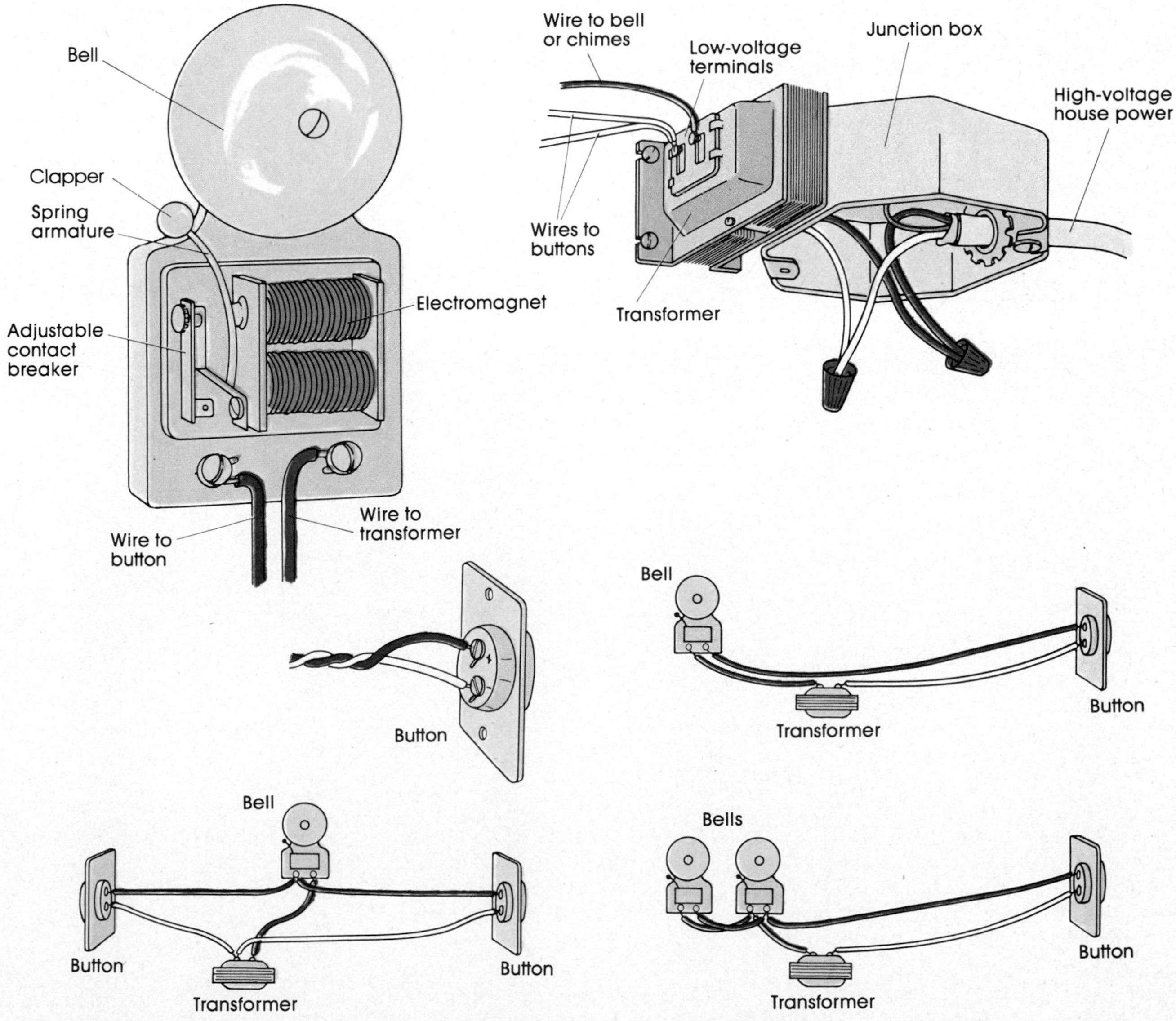

INSTALLING NEW WIRING

Any new wiring should start with your needs and a plan that meets the code. Here are the steps for installing a service entrance, a grounding electrode system, or a new circuit.

This section is designed to help you through the actual process of wiring new circuits. It takes you step by step from setting up the service entrance box to running the branch circuits through the house.

This book cannot deal with all problems you may run into while rewiring your house. If you do encounter a situation not covered here, it is recommended that you hire a licensed electrician at an hourly rate to help you on any wiring problems you cannot handle. You may have to shop around a little to find someone willing to work by the hour, but such people are available. Start by checking the classified ad section of your paper under Services. In addition to comparing their hourly rates, determine when they will be available to do the job. Hiring an electrician will cost you money, but if you know exactly what the problem is, it shouldn't take long to fix it, and you will learn something in the process.

Finally, when your wiring project is complete, it is a good practice to have an electrician go over your work before asking the building inspector for final approval. If the inspector spots some problems, it may be days or even weeks before he can return for another check.

Planning Adequate Wiring

Adequate wiring means more than wiring that just meets regulations. The NEC and local codes are concerned only with safety and do not consider what your wiring needs are, or what they might be ten years from now. So whether wiring a new home, renovating a garage, or rewiring an old home, consider your wiring needs room by room.

As time goes by, you will probably buy more appliances that draw more power. With improved wiring, you can run any appliance at any time without having to take others off the circuit or worry about blowing a

◀

You may be able to install a new circuit more easily than you think (see page 62).

fuse. In planning wiring that will meet your needs, you should first consider the following code requirements for new wiring (see the circuit diagrams on pages 26 and 27).

■ According to the NEC, you must have one circuit for each 600 square feet of habitable space, but it is better to have a circuit for each 500 square feet. You must have a minimum of three such circuits.

■ Any wall space wider than two feet must have at least one outlet. Each outlet must have another outlet within 12 feet of it, measured horizontally along the wall. The Code specifies this distance so there will be no need for extension cords. Any room dividers, such as bars or counters, are also to be included in this distance.

■ In addition to the lighting circuits, you must have the two 20-amp small-appliance circuits in the kitchen, dining room, and pantry, plus individual circuits for such appliances as the kitchen range, water heater, dishwasher, garbage disposal, washing machine, and clothes dryer. In addition, you must have an individual circuit for any permanent appliance rated at 1,000 watts or more, such as a built-in bathroom heater.

■ If you have a two-story house, make sure you have two circuits with the outlets divided between both floors. That way, a blown fuse or tripped breaker will not darken an entire floor.

■ There must be at least one receptacle in each of the following locations: behind any kitchen counter wider than 12 inches, in any room intended for laundry machines, next to each bathroom sink, in the garage, and on the house exterior.

■ The receptacles in the bathroom and garage and those outdoors must be protected by a ground fault circuit interrupter (GFCI).

■ There must be wall switches to control the lights over all outdoor entrances, hallways, and stairs, and in an attached garage. The switches at both ends of long hallways and stairs must be three-way.

SERVICE ENTRANCE

■ A wall switch must control lighting in every habitable room.

■ In kitchens and bathrooms, the wall switches must be wired to permanently installed lighting fixtures. In other rooms, they may control receptacles where lamps are plugged in.

Although the NEC does not require the following steps in wiring, many local codes do. You should incorporate them as a policy of good wiring.

If there is more than one entrance to a room, plan three-way switches at each entrance so you don't have to retrace your steps to turn a light on or off. If there are more than two entrances, then go to a four-way switch. If your house is well wired, you should be able to walk from the front to the rear turning lights on and off without having to backtrack.

There should be an adequate number of circuits for present and anticipated needs. The more circuits you have, the less chance any one will be overloaded. Although the codes do not limit the number of outlets you can put on any circuit, standard practice is to put no more than 8 to 12 receptacles and fixtures on one lighting circuit and 6 receptacles on one small-appliance circuit. The circuits should be divided equally around the house so that if a fuse does blow, one entire floor will not be darkened.

Your wiring should have enough ampacity to handle the anticipated loads in every room. In the house lighting circuit, for instance, the NEC permits No. 14 wire, which has an ampacity rating of 15 amps. But No. 12 wire is used a great deal—and even required by some local codes—because its 20-amp capacity can handle larger loads.

Installing the Service Entrance

The service entrance is where the power from the supplier, whether dropped from a pole or run underground, is connected to your house. The incoming wires pass first through the meter attached to the service box and then to the circuit breakers in the panel and to the branch circuits in the house. Virtually all service-entrance boxes are now installed with breaker switches rather than fuses. Only the breaker-switch panel is discussed here but a fuse panel is installed similarly.

Service-entrance boxes are rated in amperes, usually 100, 125, 150, and 200 amps. What you should have will be determined by the number of circuits you are planning and the computed load in your house. To determine the size of service-entrance box you need, compute the electrical load in your house.

Service-Entrance Box Components

■ **Mast.** This galvanized steel pipe, usually 2 inches in diameter, extends up the side of the house and through the eaves to rise above the roof. The mast must be high enough so that the drip loops (see illustration) are at least 10 feet aboveground and at least 3 feet from any door or window.

■ **Weatherhead.** Incoming wires are attached to this protective cover at the top of the mast. An underground service entrance will not have a mast and weatherhead, but a conduit coming up to the meter box instead.

■ **Box.** This box includes a hookup for the meter (supplied by the electric company); a neutral bus bar, where all the neutral and grounding wires are connected; and a hot bus bar, where the incoming hot wires and breaker switches are connected. In some cases, the circuit breaker

Calculating Amperage Needs

If you have or will have more than five circuits in your house, you should calculate the amperage demand to determine the size of service entrance box you will need. To calculate your amperage demands, start with the general lighting circuits by calculating the square feet in your present and planned living space, including the garage and workshop. You must supply 3 watts for every square foot of usable space.

As an example, say you have a 1,500-square-foot house. Using the Code formula of 3 watts per square foot, you will need 4,500 watts (3 × 1,500) just for the general lighting circuits. In a new house, you must have two 20-amp small appliance circuits for the kitchen. Rate each one at 1,500 watts, for an additional 3,000 watts. The laundry circuit gets an additional 1,500 watts. By checking the faceplates on other major appliances you will come up with additional wattage.

If you have just central heating, add 65 percent of its rated load in watts to the total. If you have just an air conditioner, add 100 percent of its load to the total. However, if you have a combined central air or heat, add just the larger of the two (either 100 percent of the air conditioner or 65 percent of the heater). This

is because they are not used at the same time. In the following example, we added 100 percent of the central air. The total in this example is 39,300 watts.

Kitchen range	12,000 watts
Dryer	5,000
Dishwasher	1,500
Garbage disposal	800
Water heater	5,000
Central air and heat	6,000
Lighting circuits	4,500
Small appliance circuits	3,000
Laundry circuit	1,500
Total	**39,300 watts**

Now find 40 percent of the amount over 10,000 watts. Here it is .40 × 29,300 = 11,720 watts. Add to this amount the first 10,000 watts, which brings the total to 21,720 watts. To find the amps, divide the total watts by 240 volts, which here is 90.5 amps. Thus, a 100-amp service-entrance box is sufficient, but a 125-amp box would be advisable to cover any future wiring.

Overhead and Underground Service Entrances

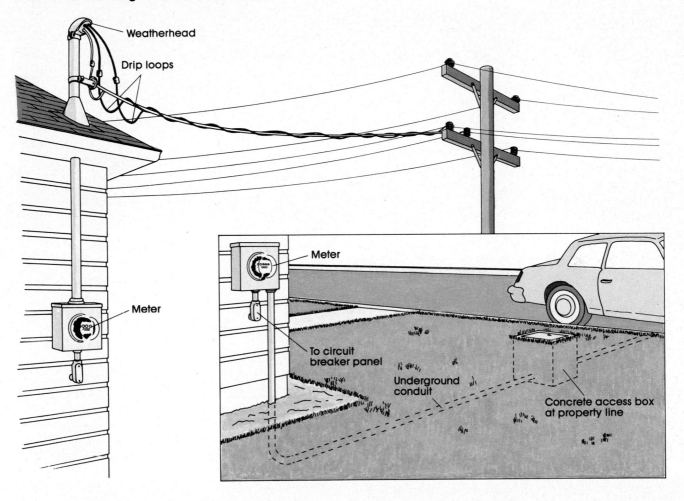

Weatherhead

Drip loops

Meter

Meter

To circuit breaker panel

Underground conduit

Concrete access box at property line

panel is located separately in the house or garage.

■ **Ground wire.** A copper ground wire is bonded (firmly attached) to the neutral bus bar and then connected with a wire-to-pipe ground clamp to the grounding electrode system.

Once you have determined the proper size of the service-entrance box and had it approved (along with your electrical project) by your building inspector, telephone your local electric company. A representative will stop by to tell you where to locate the box. You may install the service-entrance box and run all the circuits, but the electric company will not connect you with their power until you have received a final approval from the inspector. To mount the mast, weatherhead, box, and ground wire, follow these steps:

1. Mount the service box about eye level at the designated spot on the side of your house. If the house has wood siding, position the mounting screws in the studs, which are centered 16 inches apart. Locate the studs with a magnetic stud finder or by knocking on the exterior wall with a hammer until you hear a solid sound indicating the stud behind. If the walls are brick or masonry, use a concrete drill bit to drill ½-inch holes. Drive 5/8-inch wooden dowels into the holes and then put the

mounting screws in the dowels.

2. Directly above the service box, drill a hole large enough for the mast through the eaves and the roof. Drop the mast through the hole and connect it to the service-entrance box with a *hub* (water-tight nut), as illustrated. Attach the mast to the side of the house with conduit supports.

3. Attach the flashing and adjustable flashing seal to prevent rainwater from dripping through the hole and down the pipe.

4. Attach the insulator for the service drop wires. The wires may place considerable strain on the mast so make sure it is securely fastened to the house.

5. Drop the one white neutral and two black copper wires down the mast and into the service box. There can be no splices in these wires. Leave about three feet sticking out the top. Make sure they are sufficient size for the amperage requirement of your service.

6. Run the three wires through the weatherhead and attach the weatherhead to the mast. The electric company will connect your service-entrance wires to their service drop wires. You must leave enough wire sticking out—about three feet—to form a *drip loop* so that rain water will not run down the wires into the service box.

Wiring the Box

Wiring the service-entrance box may at first appear somewhat formidable, but actually it is fairly easy and straightforward. Here's the step-by-step process:

1. First wire the meter socket (as illustrated). The two hot wires are usually both black in larger sizes, but may be one black and one red. The service conductors are always attached to the top of the meter, even when they originate underground. Run the cable from the meter into the service panel.

2. Connect two more hot wires of the same size from the two screw clamps next to the lower meter sockets to the main disconnect switch.

3. Connect the neutral wire coming from the meter to the neutral bus bar in the box. Note that the neutral wire runs through the meter housing but is not electrically connected to the meter itself.

4. Connect the ground wire to the grounding electrode system of your house. The ground wire must be No. 6 or larger and should be kept as short as possible. Although not essential, encasing the ground wire in metal or plastic pipe makes your work look neater. To connect the ground wire to a pipe, use the type of wire-to-pipe ground clamp illustrated. The ground wire should be connected to the street side of the water pipe beyond the meter. Otherwise, nonmetallic fittings in the meter connection may interrupt the flow of electricity, leaving your system ungrounded. If it is not convenient to attach the wire on the street side of the water meter, connect a jumper wire around the meter, as illustrated. Use two wire-to-pipe clamps and wire the same size as the ground wire.

5. As you run each circuit through the house, connect the hot wire to the breaker switch. Connect the white neutral wire and the bare grounding wire to the neutral bus bar. If the service panel is mounted on the outside of the house, the branch circuit wires should be encased in one or more sections of conduit, from the panel to the point where they enter the house. Several circuit wires can be run in each conduit section.

A Grounding Electrode System

Until a few years ago, most residential electrical systems were grounded by simply connecting the ground wire at the service entrance to the underground water pipe. But because of the increasing use of nonmetallic water pipe and insulated fittings in metal pipes that interrupt the ground, this practice is no longer considered adequate. If you are installing a new electrical system in your house or upgrading an old one, make sure it is properly grounded by what the Code calls a *grounding electrode system*.

If the No. 6 copper ground wire on your house is now connected only to a water pipe running 10 feet or more underground, it must be supplemented by one of the following procedures to complete the grounding electrode system:

■ Connect the ground wire from the service panel to a No. 2 or larger copper wire 20 feet or longer that is buried 2½ feet deep alongside the house.

■ Connect the ground wire to 20 feet or more of ½-inch steel reinforcing rod or No. 4 or larger solid copper wire that is enclosed in concrete near the bottom of the concrete foundation of the house.

■ Connect the ground wire to the metal casing of a well (but not the drop pipe in the well).

These are important steps to take because plumbing changes around your house may have left it ungrounded without your realizing it. For example, a water meter with nonmetallic washers, which interrupt the ground, may have been installed beside the house. Or plastic pipe may have been run from the meter to your water pipes.

If your present ground is only connected to a water pipe and a water meter is less than 10 feet away, you should use a jumper wire, as shown, to go around the meter in order to continue the ground. Use two wire-to-pipe clamps and the same size wire as the ground.

In many cases, local codes also require that the ground wire be connected to the service-entrance box and to the mast for complete protection.

Service panel

No. 2 or larger wire 2½ feet deep and 20 or more feet long connected to ground wire from service panel

Ground wire connected to the well casing

Pump

Panel

Ground wire connected to ½-inch rebar or No. 4 copper wire at least 20 feet long at the bottom of the concrete foundation

Wiring a Meter Socket

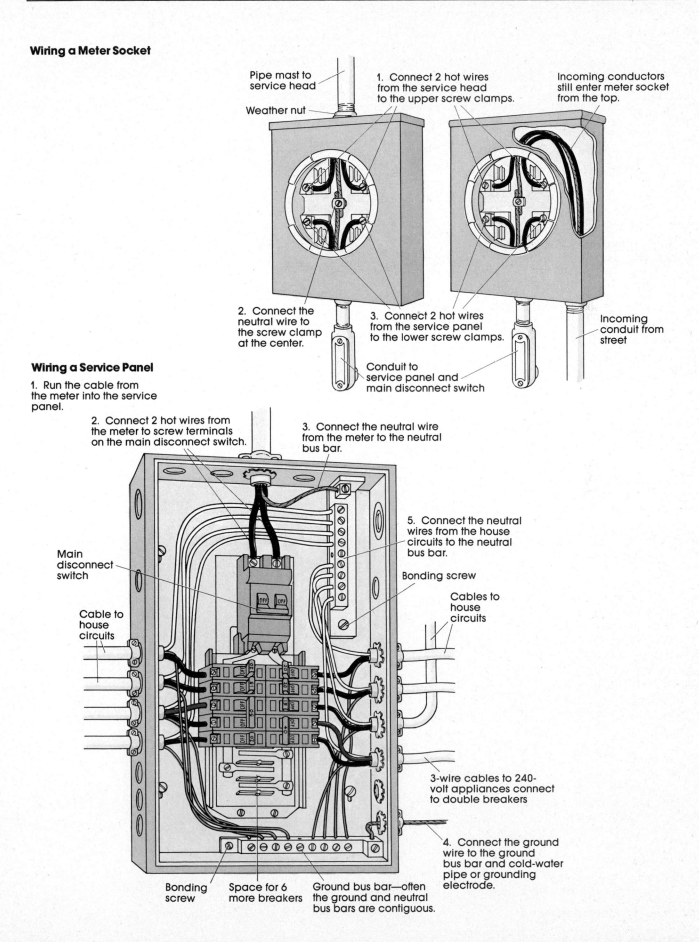

Pipe mast to service head

Weather nut

1. Connect 2 hot wires from the service head to the upper screw clamps.

Incoming conductors still enter meter socket from the top.

2. Connect the neutral wire to the screw clamp at the center.

3. Connect 2 hot wires from the service panel to the lower screw clamps.

Conduit to service panel and main disconnect switch

Incoming conduit from street

Wiring a Service Panel

1. Run the cable from the meter into the service panel.

2. Connect 2 hot wires from the meter to screw terminals on the main disconnect switch.

3. Connect the neutral wire from the meter to the neutral bus bar.

5. Connect the neutral wires from the house circuits to the neutral bus bar.

Bonding screw

Cables to house circuits

Main disconnect switch

Cable to house circuits

3-wire cables to 240-volt appliances connect to double breakers

4. Connect the ground wire to the ground bus bar and cold-water pipe or grounding electrode.

Bonding screw

Space for 6 more breakers

Ground bus bar—often the ground and neutral bus bars are contiguous.

INSTALLING A NEW CIRCUIT

Before you install a new circuit in a room or building, make a scale drawing indicating exactly where you will locate each receptacle, switch, and ceiling fixture. If three-way or four-way switches are necessary for two or more entrances, note them also. With your plan in hand, walk around the room and see where the wire must run in order to connect all boxes.

The tops of receptacles are normally located 12 to 16 inches above the floor. Determine where you want to locate the tops of the boxes; then measure and mark the spot for each box on an exposed stud. (To install boxes in finished walls, see page 69.)

The tops of wall switches are normally located 44 to 48 inches from the floor. Measure and mark each of these positions, remembering to locate them on the latch side of each door.

The nonmetallic sheathed cable that runs to each box and ceiling fixture is pulled through holes drilled in the exposed studs that form the wall. The holes should be ¾ to 1 inch in diameter to facilitate pulling the wire. Drill the holes about 24 inches above floor level. If a hole is closer than 1¼ inches to either edge of the stud, you must tack on a protective 16-gauge metal plate along the edge (see illustration). This prevents a nail from being

Wiring Plan for a New Circuit

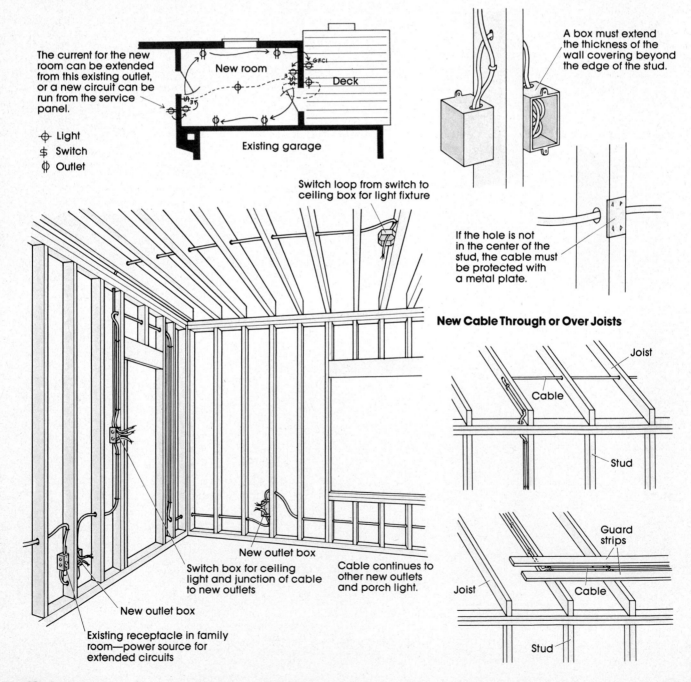

The current for the new room can be extended from this existing outlet, or a new circuit can be run from the service panel.

⊕ Light
$ Switch
⌀ Outlet

New room

Deck

GFCI

Existing garage

A box must extend the thickness of the wall covering beyond the edge of the stud.

If the hole is not in the center of the stud, the cable must be protected with a metal plate.

Switch loop from switch to ceiling box for light fixture

New outlet box

Switch box for ceiling light and junction of cable to new outlets

Cable continues to other new outlets and porch light.

New outlet box

Existing receptacle in family room—power source for extended circuits

New Cable Through or Over Joists

Joist

Cable

Stud

Guard strips

Joist

Cable

Stud

driven into the wire. However, there is no problem if you use standard 2×4 framing material and center each hole in the stud.

When you come to a door opening, bring the wire up the side of the door frame and leave an 8-inch loop where the wall switch will be located. Drill through the double top plates and the ceiling joists to carry the wire over the doorway and down the other side. Remember that unsupported wire must be stapled every 4½ feet and stapled within 12 inches of a metal box or within 8 inches of a plastic box.

If you carry wire across joists rather than running it through holes drilled in them, the wire must be protected. This is normally done by nailing two 1×1 *guard strips* along each side of the wire, as illustrated.

When drilling holes for the wire, you will not be able to fit a standard power drill between the studs. You can either drill the hole at an angle or, better, rent a special drill with a right-angle drilling head. This is what professional electricians use.

Attaching the Switch Boxes

Before you nail on the boxes for both the switches and the receptacles, you must know what type of sheathing you will have for the interior walls. The switch boxes must stick out beyond the edge of the stud an amount equal to the thickness of the wall covering so they will be flush with the finished wall. Thus, if you plan on covering the walls with ½-inch drywall, the box should protrude ½ inch from the stud.

Pulling the Wire

With all the boxes mounted, you are ready to pull the nonmetallic sheathed cable through the holes to each switch box and ceiling receptacle. Start by pushing the wire through the conduit to the service-entrance box. However, do not connect the wire to the service-entrance box until all other hookups are complete. Next pull the wire through the holes in the studs. As you come to each box, leave enough wire so that 8-inch loops of wire will stick out of the box. This will give you enough room to work freely when connecting the wires to the switches or receptacles. Avoid kinks and sharp bends, uncoil the cable carefully, and go around corners with gradual bends.

Inserting Wires in Boxes

Once you have run all the wire, cut the wires in the center of the 8-inch loops at each box. With a screwdriver, pry out the knockouts at each end of the box and insert the wires. In metal boxes, slip the wires under the clamps and tighten them down. The wire running up the stud to a metal switch box must be stapled down within 12 inches of the box. With plastic boxes, the wires must be stapled to the stud within 8 inches of the box. At every box and at the ceiling fixture, you should have two wires sticking out of the box waiting to be wired up to the switch, receptacle, or fixture. Remember that there cannot be an uncontained splice in the middle of the run. Any splice must be connected with a solderless connector, such as a wire nut, and contained in a box. Boxes that contain splices and have no other function are called junction boxes. They must have a cover plate and be accessible.

Three-Wire Appliance Circuits

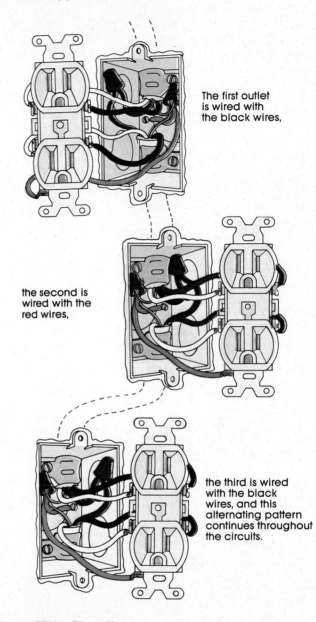

The first outlet is wired with the black wires,

the second is wired with the red wires,

the third is wired with the black wires, and this alternating pattern continues throughout the circuits.

Three-Wire Circuits

In some cases, you will have two 120-volt circuits next to each other. A good example of this is the small-appliance circuit required in new kitchens where two circuits must be run next to each other. One way to do this is to run two separate circuits. An easier way is to run one three-wire circuit.

A three-wire circuit contains a black hot wire, a red hot wire, a white neutral wire, and the bare grounding wire. Connect the receptacles just as you would in a standard two-wire system (see page 45), but with one small difference: the first receptacle is wired with the black wire, the second with the red wire, the third with the black wire, and so on. This keeps the load balanced. Make sure that the white neutral wire runs to each outlet. Never put both the red and black wire on the same receptacle since this would raise it to 240 volts.

MODERNIZING OLD WORK

By extending an existing circuit or installing a new ceiling fixture, you can change the look of a room or make it fit new uses. Be prepared for some steps new wiring doesn't involve.

Modernizing old work can mean anything from extending a circuit to rewiring the entire house. In any event, it involves cutting holes in the interior walls for switch boxes and finding a means to run the new wire from box to box. This is slow and frustrating work but it is done all the time so you can do it, too. In modernizing your old wiring, the first step is to know which outlets and switches belong to which circuit. Make a list, as explained on page 9.

Next, you should make a general inspection of your wiring wherever it is visible, such as in the basement or in the attic or crawl space above the ceiling. If you spot any errors, such as wires that are hooked up backward, or see such problems as loose wiring connections or frayed wires, then make your inspection even more thorough. Do not attempt to correct any problem you see until you are certain which circuit you are working on. Even then, double check by putting a voltage tester on a black wire and a ground and a white wire and a ground to make sure no power is coming through. If you have any doubts, shut off all power to the house. Look particularly carefully at the insulation on the old wires. Check where wires come into the fuse box and into the switch boxes. If the insulation is cracked, frayed, or worn there, it will probably be in equally bad shape throughout the rest of the house. If that is the case, you should plan on running all new wire. The first task is to remove the old wiring.

Generally, old wiring was stapled to a stud or a joist as it was run. But if it was not, as in many older houses, you can loosen the wiring at one switch box, twist it onto the new wiring, and pull the new wiring through the holes to the next box as you remove the old. However, in most cases it won't be so simple. You will need to cut off all excess old wiring from around the switch box and reroute your new wiring as described on page 62.

◄

This chapter will help you to extend a circuit or add another one in an existing room.

Old Knob and Tube Wiring

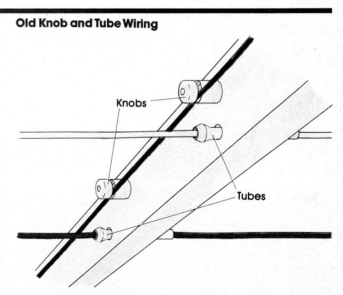

Knobs

Tubes

In older houses, you will often find a type of wiring called *knob and tube*, so named because of the porcelain support knobs for the wires and the porcelain insulating tubes pushed through holes bored in studs or joists through which the wire passed. In some cases, it was not even grounded. If the wiring in your house is not grounded, it should all be replaced.

Double check that the existing wiring is properly grounded. The ground wire from the service-entrance box must be affixed to a proper ground electrode (see box, page 60). If this is not possible, use one of the alternative grounding methods discussed on page 21.

In modernizing your old wiring, you will go through some or all of these stages:
■ extending a circuit by tapping into an existing one,
■ putting in a new circuit,
■ putting in new switch boxes,
■ pulling new wiring under the floors or through the walls.

EXTENDING A CIRCUIT

Older houses often have an insufficient number of receptacles. To add more outlets, you extend an existing circuit. Basically, that means you tie into the circuit at a certain point, determine where you want the additional outlets, then run the wire through the wall to them. The most difficult part is running the wire through the walls, which is covered on pages 62–63.

The first step is to make sure that by adding on two or three more outlets you will not overload the circuit. Check first to see whether the fuse or breaker switch is 15- or 20-amp. Then divide the total number of watts you would use on that circuit at one time by 120 volts. You must still be under 15 or 20 amps, depending on what size fuse or breaker you have. As a rule of thumb, the total number of outlets and fixtures on one circuit, including any additions, should not exceed 12.

Tapping into the Circuit

The next step is to determine where you will be able to tie into the existing circuit to extend it. Here are several possibilities:

End-of-the-run receptacle. This is about the easiest place to tie into an existing circuit. However, first use a small lamp to make sure that the outlet is not controlled by a switch. If it is, the circuit extension will not have power unless the switch is on.

With the power off on that circuit, remove the wall plates on the receptacles until you find the end of the run. You will know it because only two wires come into the box. To extend the circuit from here, attach the black wire to the brass-colored screw and the white wire to the silver-colored screw. Use a wire nut to connect the grounding wire to the existing grounding wire. Power now arrives at this receptacle and passes through it along the new wiring to the added receptacles.

Middle-of-the-run receptacle. A middle-of-the-run receptacle has four wires coming into the box, plus the grounding wires. In order to fit two more wires in there to extend the circuit, you will have to either add a box extender or *gang* two boxes together (see page 34) to make room for the additional wires.

To tie into a middle-of-the-run receptacle, you will have to join black to black, white to white, and grounding to grounding with wire nuts. Cut short lengths of each wire and strip the ends. Attach these pigtails to the appropriate screws or push-in connections and fasten them to the matching wires with the wire nuts. The incoming power is now connected to the existing receptacle by a pigtail that also links the incoming power to the circuit extension.

Middle-of-the-run switch. Use a voltage tester to determine which of the two black wires coming into the box is hot. Shut off the power and disconnect that black wire from the switch. Cut a short black pigtail and attach one end of it to the switch. Join the other end of the pigtail to the incoming hot wire and to the black wire of the circuit extension with a wire nut. Connect the white wire in the circuit extension to the existing white wire with a wire nut. Do the same with the grounding wire.

Connecting a New Wire to an Existing Circuit

End-of-the-run receptacle

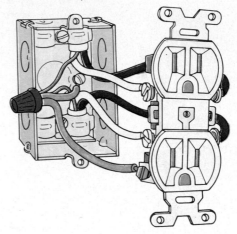

Middle-of-the-run receptacle

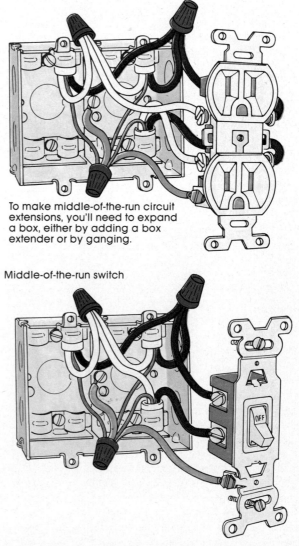

To make middle-of-the-run circuit extensions, you'll need to expand a box, either by adding a box extender or by ganging.

Middle-of-the-run switch

Installing a Junction Box

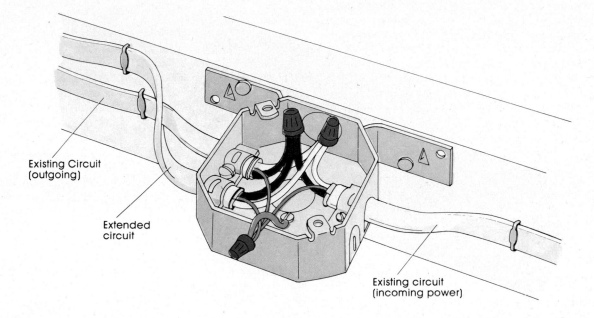

Existing Circuit
(outgoing)

Extended
circuit

Existing circuit
(incoming power)

Installing a Junction Box

Junction boxes, which are either octagonal or square, are commonly used to extend a circuit or to split one incoming power source into two or more separate circuits. Junction boxes must always be accessible and never permanently covered, such as behind walls. They must also·have a cover plate over them to protect the wiring inside.

To tap into an existing circuit with a junction box, first locate accessible wires—usually in the basement or attic. With the power off, cut the wires at the point where you will install the junction box (along a stud or over a joist). Next remove the knockouts from the junction box for the incoming and outgoing original wires plus the new wires for the extension. Fasten the junction box to the stud or joist and then slip the wires into the box. Using wire nuts, join the three black wires together, the three white wires together, and the three bare grounding wires together. To ground the box, run a pigtail from the grounding wires, connecting them to a metal screw on the metal junction box.

Tying into a Ceiling Fixture

A ceiling fixture is another convenient place to tap into when you want to extend a circuit. The ceiling fixture becomes, in essence, a junction box. With the power off, open the ceiling fixture to expose the wiring. You will see an incoming white wire attached to the fixture, and an incoming black wire attached to the white wire that leads to the switch. This system indicates a switch loop (see page 44), and the white wire should be painted black or have black tape on it to indicate that it is hot. The bare grounding wires are attached to each other and a pigtail to the metal screw on the fixture box.

To extend the circuit, cut the white wire leading to the fixture and then use a wire nut to connect the cut wires with the white wire on your circuit extension. In the same manner, join the circuit extension black wire to the connection between the incoming black wire and the hot white wire (marked black) that leads to the switch. Finally, connect the circuit extension grounding wire to the other grounding wires.

Do not tie into a ceiling fixture with only two wires coming into it. This indicates it is at the end of a switch-controlled run and will have no power when the switch is off.

Tying into a Ceiling Fixture

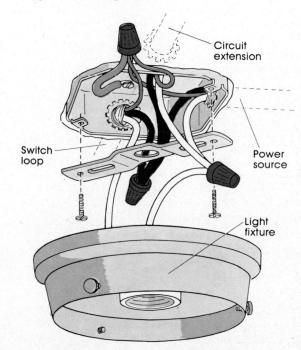

Circuit
extension

Switch
loop

Power
source

Light
fixture

Installing a Raceway or Plug-in Strip

The easiest way to extend wiring in your house is to use plug-in strips or raceways. These are quickly mounted on the surface of a wall without elaborate fishing of cable through walls. Because they are visible, some people consider them less attractive than standard wiring, but it is a quick way to gain several additional outlets. While they can be used in any room, raceways and plug-in strips are particularly useful along kitchen counters or above the bench in a workshop. They are, in effect, an extension cord that fits neatly on the wall. Just be sure that you don't abuse this extended circuit by using too many appliances on it at the same time.

Plug-in strips. The easier of these two wiring methods to install is the plug-in strip, a rigid plastic wiring strip that plugs into specially made switches or receptacles. One type can be easily attached to walls or baseboards, and another type is wired to the receptacles or switches at the terminal screws.

To install the strips, simply lay out the number of receptacles or switches you want along the wall or baseboard. Turn off power to the existing receptacle, which is the power source, then plug in the wire and run it to the first new receptacle. The strips are attached to the wall and baseboard in different manners, according to the manufacturer's instructions. Plug in or wire the strips to the first receptacle or switch and continue along the wall to the other devices. Turn the circuit back on and test your work by plugging in a lamp.

Metal raceways. Some metal raceways are designed to match the height and thickness of a baseboard. You simply remove a section of baseboard, install the raceway in its place, and paint the raceway to match the baseboards. Other metal raceways have narrow metal strips that fit inconspicuously on the wall just above the baseboard.

To install the first type, remove an amount of baseboard equivalent to the length of the raceway. If possible, start one end directly under an existing outlet, which will be the power source. If this is not practical, you can run cable from the nearest receptacle through the wall to the raceway. Pull the new wires from the raceway that are connected to an existing receptacle (in a standard fashion) through the predrilled hole in the end of the raceway and hold them in place there with the cable clamp provided. In many models, the raceway is already wired. Use two wire nuts to connect the incoming wires to the raceway wires—black to black, white to white. Screw the raceway in position on the wall and snap on the front cover.

In the second type of raceway, which uses narrow metal strips positioned just above the baseboard, no drilling into the wall or removal of baseboard is required. To wire the raceway into an existing receptacle, first shut off power on that circuit. Remove the wall plate and the receptacle but leave the wires intact. Install the provided extender box over the existing receptacle according to the directions on your particular model. This extender allows you to wire the raceway into the receptacle without drilling into the wall. Measure and cut the raceway according to your needs and screw the raceway backing to the wall. Connect the wires to the receptacles as you would standard receptacles. Then snap the cover plate on the raceway. Turn the circuit back on and check your work.

Raceways and Plug-In Strips

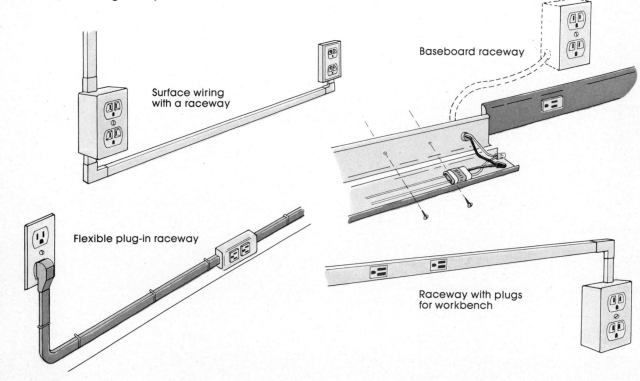

Surface wiring with a raceway

Baseboard raceway

Flexible plug-in raceway

Raceway with plugs for workbench

INSTALLING SWITCH BOXES IN WALLS

Switch boxes can be installed in three basic ways: nailed to a stud, clamped in the wall between the studs, or screwed to lath behind the plaster. Several different types of boxes are illustrated and explained on page 34. Choose the one that best suits your need.

Where to Locate the Switch-Box Opening

New switch-box openings should be kept the same height as the existing ones, whether for receptacles or for switches. In a new room, locate the switches from 44 to 48 inches from the floor and the receptacles from 12 to 18 inches from the floor.

If you plan to mount the switch box on a stud behind the wall, you must first locate the stud. If you are going to put the box in the wall between the studs, you still must locate the stud so it will not be in the way. To find the stud, rap on the wall sharply with your knuckle. Go back and forth until you hear a solid sound, as opposed to the hollow sound between the studs. If that does not seem satisfactory, buy a *stud finder* at a hardware store. It has a magnetic needle that swings into line when the finder passes over nails holding the wall sheathing to the stud. An even more sure way to find the stud is to drill a ⅛-inch test hole where you plan to put the box. If it is over a stud, you will see wood shavings from the stud come out of the test hole.

If you want the box between the studs, drill the test hole where you plan to mount the box. Then take an 8-inch-long piece of wire, bend it in half at right angles, and insert it in the hole. Rotate the wire in all directions to make sure you do not run into obstructions. If you do, move over 4 to 6 inches. Insulation in your walls can make this difficult to do.

For lath-and-plaster walls, drill the test hole and use the wire test. Next chip away enough plaster around it to see if you are in the middle of one of the laths, which is where the box should be centered. You may chip away a little more plaster, either down or up, until the opening is centered over a lath. Work carefully with the chisel in order not to crack the surrounding plaster.

Installing a Box in Plasterboard

When you have determined exactly where you want to locate the box, which should be designed for location between studs in drywall, place the open side of the box against the wall and trace its outline. Cut along the outline with a sharp pocket knife or utility knife. Use a straightedge to guide the cut. You do not need to cut through the drywall itself, but you must cut completely through the paper. You may have to go around the outline two or three times.

Next place the end of a short length of 2 × 4 in the center of the outline and give it a sharp rap with a hammer. Once you have broken through the plasterboard, trim away all the ragged edges along the outline with your knife. The plasterboard is sandwiched between two layers of paper, and you must cut the inside layer of paper to remove the plasterboard fragments.

Pull the wires through the opening so that 8 inches are sticking out. Remove the knockouts from the box for

Plasterboard Walls

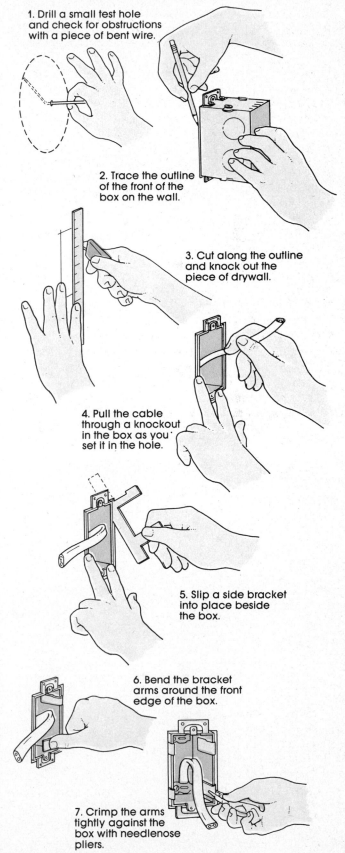

1. Drill a small test hole and check for obstructions with a piece of bent wire.

2. Trace the outline of the front of the box on the wall.

3. Cut along the outline and knock out the piece of drywall.

4. Pull the cable through a knockout in the box as you set it in the hole.

5. Slip a side bracket into place beside the box.

6. Bend the bracket arms around the front edge of the box.

7. Crimp the arms tightly against the box with needlenose pliers.

the wire to travel in the most direct path. As you insert the box into the opening, pull the wires through the knockout holes. Don't tighten the wire clamps in the box until the box is completely installed.

To install the box, first adjust the top and bottom ears so that the front of the box is flush with the wall. Next, while holding the box in place with one hand, slip one side bracket between the wall and the box. Insert the top of the bracket first, push it up far enough to slip the bottom of the bracket into the opening, pull the bracket snug against the inside of the plasterboard, and bend the tabs over to clamp the box. Do the same on the other side. If you drop the second bracket behind the wall, you will have to remove the first bracket and the box and fish it out, so be careful.

When the box is installed, use a pair of needlenose pliers to firmly squeeze the tabs against the box. Pull the wires until about 8 inches extends out the box and then tighten the clamps.

Installing a Box in Plaster Walls

Determine where you will locate the box and then drill a small test hole to insert wire to make sure you are not too close to a stud (as detailed on page 69). Using a cold chisel and hammer, chip away a small section of plaster until you expose a lath board behind the plaster. In some cases, the backing may be a metal screen, called *metal lath.*

If the backing behind the plaster is a metal lath, use a standard box rather than one with ears. Place strips of

Metal Lath and Plaster Walls

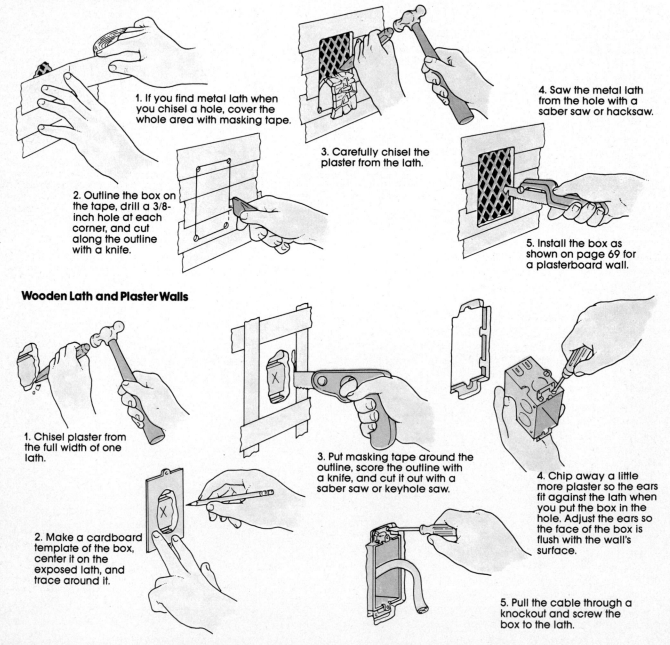

1. If you find metal lath when you chisel a hole, cover the whole area with masking tape.

2. Outline the box on the tape, drill a 3/8-inch hole at each corner, and cut along the outline with a knife.

3. Carefully chisel the plaster from the lath.

4. Saw the metal lath from the hole with a saber saw or hacksaw.

5. Install the box as shown on page 69 for a plasterboard wall.

Wooden Lath and Plaster Walls

1. Chisel plaster from the full width of one lath.

2. Make a cardboard template of the box, center it on the exposed lath, and trace around it.

3. Put masking tape around the outline, score the outline with a knife, and cut it out with a saber saw or keyhole saw.

4. Chip away a little more plaster so the ears fit against the lath when you put the box in the hole. Adjust the ears so the face of the box is flush with the wall's surface.

5. Pull the cable through a knockout and screw the box to the lath.

wide masking tape on the wall and trace the outline of the box over the tape. With a ⅜-inch metal bit, drill holes at the four corners. Score the outline several times with a sharp knife, and then use a cold chisel to remove the plaster outside the outline. To cut the metal lath once the plaster is removed, use a metal cutting blade in a saber saw. Starting in one corner hole, work your way around the outline. Work slowly so the saw vibration will not crack the plaster.

If the backing is not metal lath, once you have exposed the full width of a single lath, mark the center with a small "X." Next, cut a template for the box by placing it face down on a piece of cardboard, tracing its outline, and cutting it out. Cut a small hole in the center of the template and then center that hole over the "X" on the lath. Trace the outline of the template on the wall. Cover the outline with wide strips of masking tape to help prevent the plaster from cracking. Retrace the template outline over the masking tape.

Score the outline several times with a sharp knife and then drill ⅜-inch holes at each corner and at the curves in the top and bottom. Cut along the outline with a keyhole saw or saber saw. Note that when you are finished you will have cut away the center lath board and half of the top and bottom ones.

Place the box in the hole, outline the ears, and chip away the plaster there so the ears will fit flush against the lath boards. Work slowly and carefully to minimize damage to the surrounding plaster. Position the moveable ears so that the box is flush with the plaster. Then drill pilot holes in the lath for the mounting screws. Pull the wire through the knockouts, install the box, and tighten the clamps in the box to hold the wire.

Installing a Box in Wood Walls

Before you cut into a wood wall, drill a small test hole and insert a piece of bent wire (as described on page 69) to make sure you are not too close to a stud or pipe. Place the box face down on the wall and trace its outline with a pencil. Do not trace the ears but go straight across the top and bottom. If the wall is ⅜ inch or thinner, use a box with adjustable side clamps. If the wall is thicker than ⅜ inch, use a standard box without the side clamps.

With a ⅜-inch bit, drill holes at the four corners. In addition, drill a hole at the top and bottom centers to provide clearance for the long screws that fasten the switch or receptacle to the box. With a keyhole saw, cut along the outline. Once that is finished, place the box in the hole and adjust the ears on top and bottom so the box is flush with the wall surface. Remove the box, pull the wires through the knockout holes, put the box back in place, and drill pilot holes for the mounting screws at the top and bottom. The wall plate will cover the ears.

If the wall is ⅜ inch or thinner, and you must use a box with side clamps, drill a hole inside the bulges on each side of the traced outline as a starting point for the saw cutting the curve. Once the box is installed and the wires are pulled through, tighten the clamps on the wire in the box.

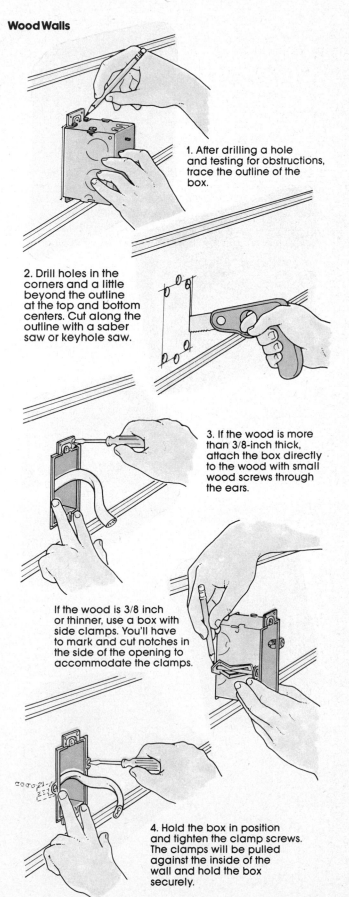

Wood Walls

1. After drilling a hole and testing for obstructions, trace the outline of the box.

2. Drill holes in the corners and a little beyond the outline at the top and bottom centers. Cut along the outline with a saber saw or keyhole saw.

3. If the wood is more than 3/8-inch thick, attach the box directly to the wood with small wood screws through the ears.

If the wood is 3/8 inch or thinner, use a box with side clamps. You'll have to mark and cut notches in the side of the opening to accommodate the clamps.

4. Hold the box in position and tighten the clamp screws. The clamps will be pulled against the inside of the wall and hold the box securely.

INSTALLING CEILING BOXES

Of the three basic types of ceiling boxes, which one you choose will depend on two factors. First, you must know whether your ceiling is made of plasterboard or, as in older houses, of lath and plaster. Second, you must consider whether you have room to work in an attic or crawl space above the ceiling or whether you must work beneath the ceiling.

Bar hanger box. This is about the most popular type of hanger if you have access above the ceiling. The box is fastened to the bar hanger by a *stud screw* through the stud (which supports the lighting fixture). The bar hanger can be adjusted to fit between the ceiling joists, and the box can be slid back and forth on the bar until it is positioned just where you want it in the ceiling.

Flange box. This box is easy to install from above since the flange is simply nailed or screwed to the ceiling joist. The chief limitation is that it must be fastened to a ceiling joist and that joist may not be where you want to locate the light. However, you can put a cross piece between two joists and then nail the flange box to that.

Offset hanger box. This box is commonly used in lath-and-plaster ceilings where you must work from below the ceiling. The hanger fits in a channel cut through the ceiling plaster between two joists.

Lightweight fixture boxes. In addition to the three main types of ceiling boxes, two are designed for lightweight fixtures. The *ceiling cut-in box* is convenient for fixtures weighing less than two pounds. Once installed from below with the spring-flanges spread, it is very difficult to remove. With the shallow *pancake box*, you need not cut the ceiling but just screw the box to the bottom of a joist through the ceiling material, and then drill holes through the cutouts for the wires. It will accommodate only five wires.

Ceiling Boxes

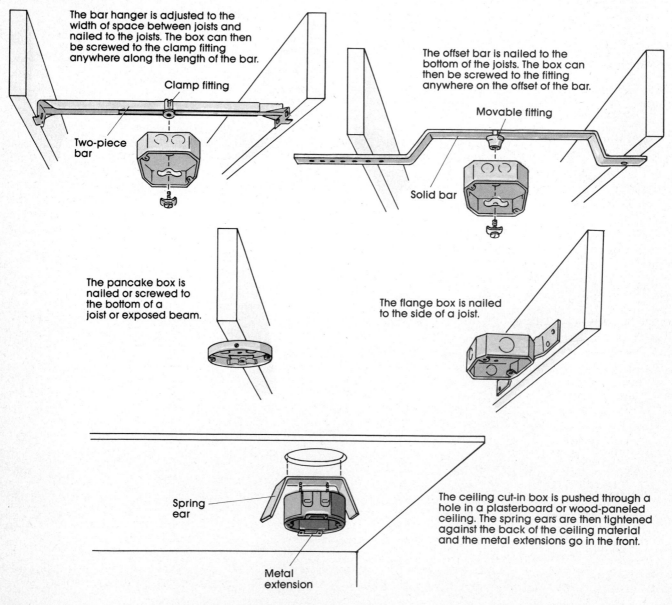

The bar hanger is adjusted to the width of space between joists and nailed to the joists. The box can then be screwed to the clamp fitting anywhere along the length of the bar.

Clamp fitting

Two-piece bar

The offset bar is nailed to the bottom of the joists. The box can then be screwed to the fitting anywhere on the offset of the bar.

Movable fitting

Solid bar

The pancake box is nailed or screwed to the bottom of a joist or exposed beam.

The flange box is nailed to the side of a joist.

Spring ear

Metal extension

The ceiling cut-in box is pushed through a hole in a plasterboard or wood-paneled ceiling. The spring ears are then tightened against the back of the ceiling material and the metal extensions go in the front.

Installing Ceiling Boxes from Above

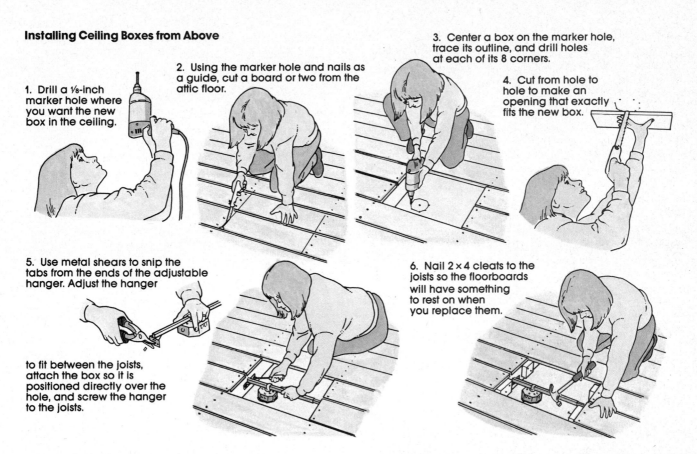

1. Drill a ⅛-inch marker hole where you want the new box in the ceiling.

2. Using the marker hole and nails as a guide, cut a board or two from the attic floor.

3. Center a box on the marker hole, trace its outline, and drill holes at each of its 8 corners.

4. Cut from hole to hole to make an opening that exactly fits the new box.

5. Use metal shears to snip the tabs from the ends of the adjustable hanger. Adjust the hanger to fit between the joists, attach the box so it is positioned directly over the hole, and screw the hanger to the joists.

6. Nail 2 × 4 cleats to the joists so the floorboards will have something to rest on when you replace them.

Working with Access from Above

The simplest type of installation involves a plasterboard ceiling with working room above it and ceiling joists not covered by flooring. In this case, put down some boards over the joists so you will not accidentally step on the ceiling, which would break easily under your weight. If there are floorboards, you can cut out two or three of them and then replace them after the ceiling box is installed. Here are the steps to follow:

1. Start from below by determining where you want the box to be located in the ceiling. Mark that spot and drill a hole in the center with a ⅛-inch bit. You should break through either plasterboard or lath and plaster almost immediately since both are normally 1-inch thick or less. If you see wood shavings from a plasterboard ceiling, you have hit a joist. Move about 3 inches to one side of the joist and drill another hole. You can patch the first hole with spackling compound, available in any hardware store. In a lath-and-plaster ceiling, you will probably see some wood shavings as you drill through the laths. If there are floorboards in the attic, put an 18-inch extension on the drill and use a ⅛-inch bit to drill a pilot hole through the floorboards after drilling through the ceiling.

2. Working in the attic, locate the hole in the floor and prepare to remove that section of board between the joists. You can easily spot the ceiling joists because the flooring is nailed over each joist. From the nails, measure toward the hole about 1 inch. This will locate the edge of the joist. Then use a square to draw a straight line across the floorboard. Drill ⅜-inch holes on the line at one edge of the floorboard and then use a saber saw or keyhole saw to cut the board at each end. Save the board to replace it when the project is finished.

3. Center the ceiling box over the marker hole drilled through the ceiling. If it is 4 inches or less from a joist, you must use a flange box that attaches directly to the joist. If the box is more than 4 inches from a joist, use a bar hanger. In either case, position the box face down and trace its outline on the plasterboard. Use a ⅜-inch bit to drill a hole at each of the eight corners of the box.

4. Working from under the ceiling again, use a saber saw or keyhole saw and cut from hole to hole. If the ceiling is lath and plaster, place strips of wide masking tape along the area to be cut to help keep the plaster from cracking. The ceiling can also be braced against the saw vibrations by holding a 1 × 6 board about 2 feet long against the ceiling next to the cutting line.

5. To install the box, first remove any knockouts necessary to run the wires. Fasten the box to the hanger with the stud and stud screw. Use metal shears to snip off the tabs at each end of the hanger, then position the bar hanger so the box is just flush with the ceiling surface. With a pencil, mark the joist through the screw holes in the hanger at each end. Then remove the hanger and drill the pilot holes for the screws. Reposition the hanger and screw it to the joists. If you are using a flange box, position it over the hole, mark the mounting holes, drill the pilot holes, and then screw the box to the joists.

6. When the ceiling box is mounted, nail a 2 × 4 cleat to each joist just under the floorboards. Then put the floorboard section back in position and nail it to the cleats.

Working from Below with Plasterboard

In many instances, it is impossible to install a new ceiling box from above the ceiling because there is no working room. If this is the case, you must remove part of the ceiling in order to gain access to the ceiling joists. To do this, you must remove a whole section of the plasterboard between two joists. It may seem easier to just cut a small piece out and patch it up. But putting in a large piece of plasterboard is just about as easy as trying to patch a small hole and you have more room in which to work. Here's how to install a ceiling box from below:

1. Determine where you want to locate the ceiling box. Drill a ⅜-inch hole and insert a bent piece of wire to make sure the area is free of pipes or other obstructions. Starting from the small hole, use a keyhole saw to make a square hole about 8 inches across.

2. Insert a steel tape at one corner of the opening and measure the distance to a joist. Now measure that same distance on the surface of the ceiling and add ¾ inch to mark the center of the joist. Move to the other corner of the opening on the same side and repeat the process.

This shows the exact position of the ceiling joist. Now repeat that process on the other side of the square to locate the center of the other joist. With a steel square as a guide, draw a line through the two marks under each joist. You now have two parallel lines either 16 or 24 inches apart, depending on the spacing of your ceiling joists. With these two lines on the outside perimeter, use the steel square to draw either a 16-inch or 24-inch square on the ceiling.

3. Using the steel square as a guide, cut the plasterboard along the drawn lines. Be careful not to cut beyond the corners. You do not need to cut through the plaster, but you must cut through the exterior paper layer. After scoring the edges of the square, cut two diagonal lines across the square. This makes it easier to break out. With a hammer, break out the plaster inside the square. Cut the paper backing away along the edges with your knife. Pull the nails from the joist and use a chisel to chip away any plasterboard stuck to the joists.

4. Cut a new piece of plasterboard to the precise dimensions of the hole in the ceiling *less* ⅛ inch on all sides.

Installing a Ceiling Box in Plasterboard

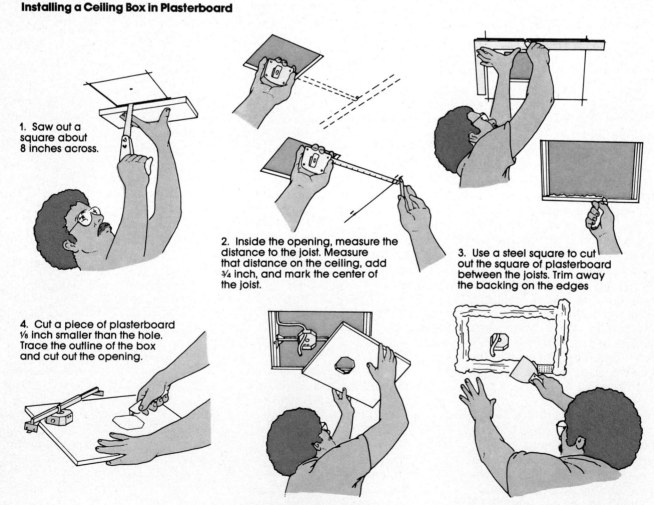

1. Saw out a square about 8 inches across.

2. Inside the opening, measure the distance to the joist. Measure that distance on the ceiling, add ¾ inch, and mark the center of the joist.

3. Use a steel square to cut out the square of plasterboard between the joists. Trim away the backing on the edges

4. Cut a piece of plasterboard ⅛ inch smaller than the hole. Trace the outline of the box and cut out the opening.

5. Tack the hanger and box in place and nail the ceiling panel along the edges.

6. Press plasterboard joint compound around the panel, apply joint paper, and spread 4 thin coats of compound on top.

Plasterboard can be cut by first drawing the dimensions then scoring the lines with a knife deeply enough to cut through the exterior layer of paper. Place the cut line just over the edge of a table or workbench and hit that section with the palm of your hand. After the board has snapped along the scored line, cut the paper backing. Place the ceiling box face down in the center of the new panel and trace its outline. Cut along the outline with a knife deeply enough to slice through the paper. Put the panel on four blocks of wood placed around the edge of the box outline and break out the box opening with a hammer. Cut away the paper backing.

5. Tack the bar hanger and ceiling box in place between the joists and then hold the ceiling panel up to check that the box is centered in the panel hole. After any necessary adjustments, screw the bar hanger to the joists. Run the wires into the ceiling box and tighten the clamps. Put the ceiling panel in place and nail along the edges every 4 inches.

6. With a putty knife, press plasterboard joint compound into the cracks around the new panel. Then apply strips of perforated joint paper. Do not overlap at the corners. Use the putty knife to press the paper into the compound and smooth it out. Spread a thin layer of compound over the tape, keeping it as smooth as possible. Let it dry and then apply three more coats. Sand the final coat smooth and paint to match your ceiling.

Working from Below in Lath and Plaster

Installing a ceiling box in lath and plaster is arduous if you are working from below. A lot of plaster falls in your face as you work so it is advisable to wear protective goggles. For this job you basically need a hammer, and a 1-inch cold chisel. Here's the step-by-step process:

1. Since the offset bar hanger will hang between two joists, the first step is to locate the joists. Starting from the point where you want to position the ceiling box, chip a small hole until you uncover a lath. Chisel along that lath until you uncover a nail through the lath, indicating the point where it is nailed to the joist. Do the same in the opposite direction, making the channel as wide as a single lath, which will be removed. Place wide strips of masking tape along each side of the channel to help prevent the surrounding plaster from cracking.

2. Trace the outline of the ceiling box where you want it and then drill a ⅜-inch hole at each corner of the box outline. Cover the surrounding area with masking tape and then use a keyhole saw to cut from hole to hole along the lines. Hold a 2-foot length of 1 × 6 or piece of plywood next to the line you are cutting to keep the saw from pulling the surrounding plaster loose.

3. With a keyhole saw, cut the exposed lath on the outside of the two joists and remove the nails. Drill pilot holes for the screws to secure the bar to the joists. Attach the box to the hanger, pull wires through the box, and then screw the hanger to the joists.

4. Fill the channel with patching to a level about ⅛ inch below the ceiling surface. When the plaster has set, smooth in a layer of spackling compound. Once that has dried, sand it smooth and paint it.

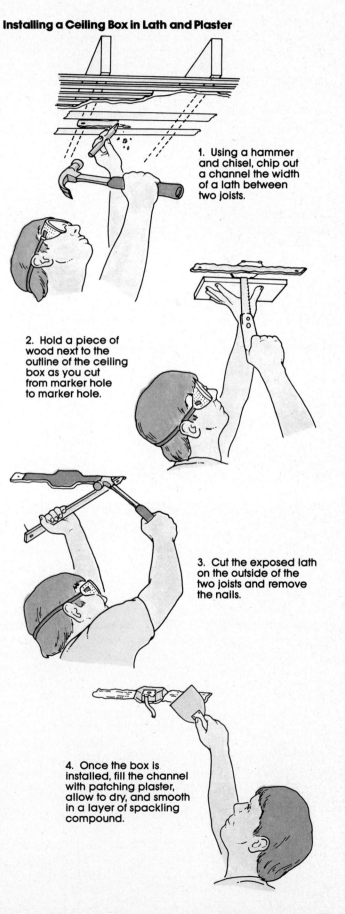

Installing a Ceiling Box in Lath and Plaster

1. Using a hammer and chisel, chip out a channel the width of a lath between two joists.

2. Hold a piece of wood next to the outline of the ceiling box as you cut from marker hole to marker hole.

3. Cut the exposed lath on the outside of the two joists and remove the nails.

4. Once the box is installed, fill the channel with patching plaster, allow to dry, and smooth in a layer of spackling compound.

To Run Cable Under the House

1. Below the end-of-the-run receptacle, drill a pilot hole through the floor and push a piece of wire through it.

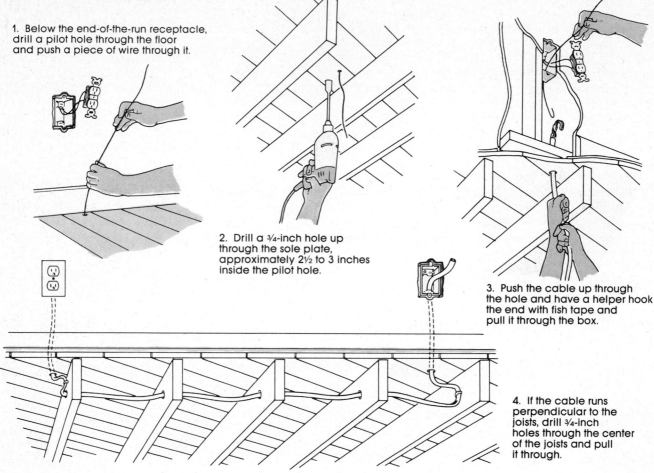

2. Drill a ¾-inch hole up through the sole plate, approximately 2½ to 3 inches inside the pilot hole.

3. Push the cable up through the hole and have a helper hook the end with fish tape and pull it through the box.

4. If the cable runs perpendicular to the joists, drill ¾-inch holes through the center of the joists and pull it through.

Running new wire for additional outlets or switches in a finished house can be fairly simple or quite tedious. Two people, both with considerable patience, are normally required for this work.

One of the difficulties of the job is that the wall studs, through which the wiring is normally run, are covered. The easiest way to get around this problem is to run the additional wiring under the house or through the attic space above the house. When that isn't possible, the wire can be run behind the baseboard in a room, or even through the studs, by removing a section of the plasterboard wall.

You will need a few special tools for running additional wire in a complete house. Because you have to fish for the wire in places you can't see or reach with your hands, you will need *fish tape*, a 25-foot roll of spring steel or stiff nylon that can be pushed through small holes and down into wall spaces (see page 10). One end is bent into a hook and is used to catch and hold the hooked end of new wiring that is pushed through holes drilled behind the walls. A *power drill* with several different bits is also essential. You will need a ¹⁄₁₆-inch bit for drilling pilot holes and a ¾-inch space bit for drilling holes through joists and plates. Depending on your house construction, you may need an 18-inch extension bit. *Cable staples* are required to support the cable. The cable must be

stapled within 12 inches of a new outlet or switch and every 4 feet along joists or studs.

Running Cable Under the House

In many older houses, there aren't enough outlets per room. The easiest way to add a few more outlets is by running new cable under the house. If you have a basement or crawl space under the house, you can do the job. Here's how:

1. Remove the wall plates on receptacles in the room to locate the end-of-the-run receptacle, which will have only two wires coming into the box. That is the box you are going to tie into with the additional wiring. Directly in front of this box and right next to the baseboard, drill a ¹⁄₁₆-inch pilot hole down through the floor. Push a long piece of wire through the hole so you can quickly locate it when you are under the house.

2. Working under the house, directly in line with the pilot hole, drill a ¾-inch hole up through the *sole plate* (bottom of stud wall). A spade bit does a fast, clean job. After making sure that no power is coming into the outlet, remove it and pull it out of the box. Leave the wires on. Punch out the knockout in the bottom of the box.

3. Strip about 4 inches of insulation from the end of the cable and bend the wires into a hook. Push the cable through the hole in the plate up toward the outlet box. Have a helper push a fish tape (for this short distance you could use a wire coat hanger) through the knockout hole in the box. When the helper hooks the cable, pull it out of the box about 12 inches. Tighten the clamp in the box on the cable.

4. Staple the wire to the joist to keep it in place and then run the wire under the house to the next receptacle location. If the wire runs parallel to a floor joist, staple it every 4 feet to the side of the joist. If the cable runs perpendicular to the joists, drill ¾-inch holes in line through the center of the joists and pull the wire through. At the new receptacle locations, cut a hole in the wall for the box and mount the box, as shown on page 69. Drill a pilot hole directly under the box and repeat steps one and two to get the wire into the box.

Some local codes require new cable under the house to be contained in thinwall conduit. Check with your local building inspector.

Running Cable Through the Attic

If you don't have room to crawl under the house but do have an attic, you can run the wire above the ceiling from an existing outlet to a new one. The process is essentially the same as going under the house, except that you are working from above. Complete steps one through four for running a cable under the house, with the following differences: Drill the ¹⁄₁₆-inch pilot hole through the ceiling directly above the existing outlet (see step 1). Drill the ¾-inch hole down through the top plates next to the pilot hole (see step 2). When the helper pushes the fish tape down to box level, hook it with a piece of bent coat hanger or wire. Pull the fish tape down through the box and then hook it to the stripped end of the new cable. Tape the two together so they won't slip apart, then have your partner pull the wire up through the wall and over to the new ceiling box or to where you will run the cable down the wall by the same process to a new outlet (see step 3).

To Run Cable Through the Attic

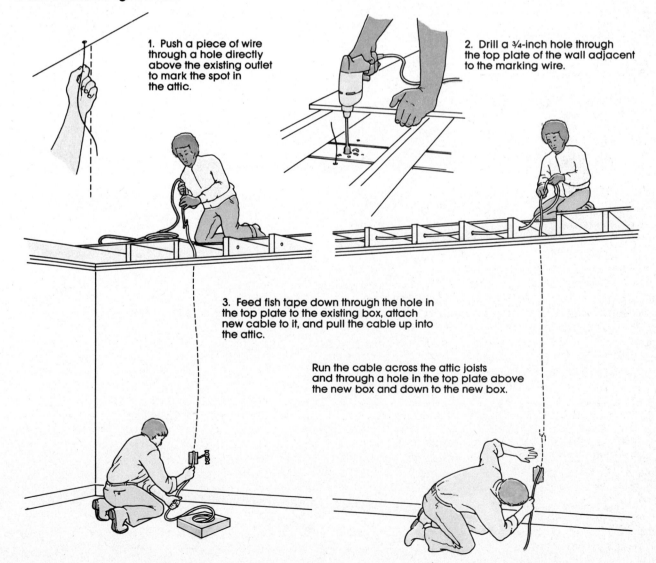

1. Push a piece of wire through a hole directly above the existing outlet to mark the spot in the attic.

2. Drill a ¾-inch hole through the top plate of the wall adjacent to the marking wire.

3. Feed fish tape down through the hole in the top plate to the existing box, attach new cable to it, and pull the cable up into the attic.

Run the cable across the attic joists and through a hole in the top plate above the new box and down to the new box.

Routing Cable Behind a Baseboard

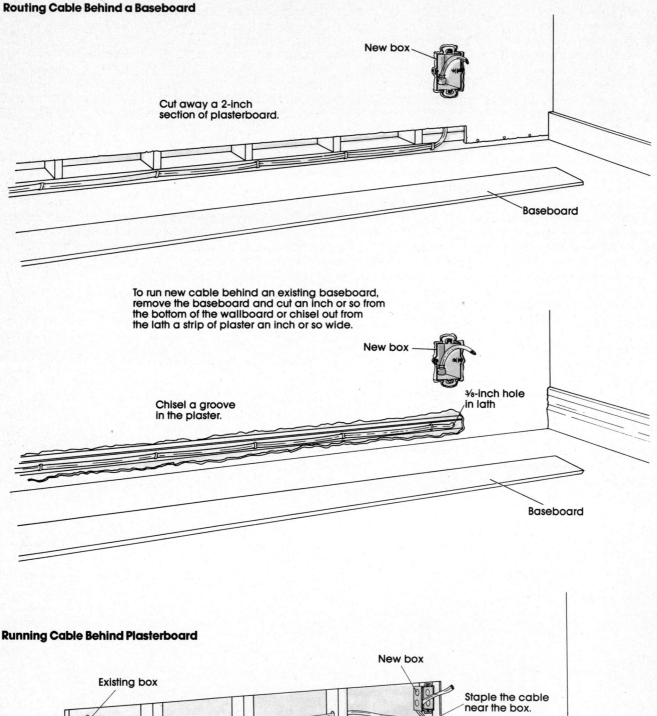

New box

Cut away a 2-inch
section of plasterboard.

Baseboard

To run new cable behind an existing baseboard,
remove the baseboard and cut an inch or so from
the bottom of the wallboard or chisel out from
the lath a strip of plaster an inch or so wide.

New box

⅜-inch hole
in lath

Chisel a groove
in the plaster.

Baseboard

Running Cable Behind Plasterboard

New box

Existing box

Staple the cable
near the box.

To run cable through studs behind existing wallboard,
remove a 6-inch-wide section of the wallboard from
the center of the studs nearest the existing box to
the new box.

Routing Wire Behind a Baseboard

When you want to add one or more receptacles along one wall, it can often be done readily by removing the baseboard and running the wire in the wall behind it. This is particularly true in lath-and-plaster houses, which often have an 8-inch-high baseboard. This wiring method can be used in either a plasterboard or lath-and-plaster wall. Here are the steps involved:

1. Remove the baseboard. Using a stiff putty knife, first go along the top of the board and cut through the layer of paint between the wall and the baseboard. With a thin chisel, carefully pry the board away from the wall. Work slowly and keep using the chisel rather than just pulling on one end of the loose board. Doing that may crack the baseboard.

2. Make sure the electricity is off and remove the receptacle from the existing box, leaving the wires intact. Punch out the knockout in the bottom of the box. Cut a hole in the wall for the new receptacle (see page 70).

3. In lath and plaster, chisel a channel in the plaster to the depth of the studs from the existing box to the new one. In plasterboard, remove a 2-inch-wide section of the plasterboard between the two boxes.

4. Drill a ⅜-inch hole in the wall at each end of the channel beneath the two boxes. Fish a wire up to the existing box, run the wire down the channel, and fish that end up to the new box. Staple the wire to each stud.

5. Install the new box and receptacle.

6. Replace the baseboard, being careful not to drive nails where the wire is located.

Running Cable Behind Plasterboard Walls

The walls in most houses across the country are made of 4 × 8 plasterboard sheets. Here is the easiest way to run new cable behind these walls:

1. Make sure the power is off at the existing receptacle. Remove the wall plate and receptacle from the box. With a screwdriver, punch out the bottom knockout in the box.

2. Starting from the center of the stud that holds the existing receptacle, remove a 6-inch-wide piece of plasterboard across the wall to the location for the outlet. Stop in the middle of the stud where the new box will be fastened. Measure and cut this section of plasterboard to be removed very precisely so that the replacement piece can be cut and smoothly fitted into the opening.

3. Drill ¾-inch holes in all the studs between the boxes. Run the cable through the holes and hook it up to the receptacles.

4. Cut a piece of plasterboard ⅛-inch smaller all around than the hole across from the old receptacle to the new. Use two nails wherever it crosses a stud. Drive the nails hard enough to dimple the plasterboard. Then cover the nail heads with plasterboard joint compound. Fill the crack with joint compound, press in strips of perforated joint tape, and smooth with a broad putty knife. Smooth a thin layer of compound over the tape and let dry. Apply three more layers of compound, sanding the final one smooth, and then paint.

Running Cable Around a Doorway

When installing new wiring, you may have to route it around a doorway in order to reach another wall. The task is not as formidable as it might appear. Here's how to do it:

1. Remove the molding or casing around the outside of the door frame. Use a putty knife to break the paint bond and then slowly pry the boards loose with a broad, thin chisel. Work carefully to prevent cracking the molding.

2. Run the cable from the existing outlet to the doorway either behind the baseboard or by cutting out a section of the plasterboard wall.

3. Run the cable up between the stud and the door jamb. Wooden spacers are often used between the stud and jamb to align the doorway. Just notch them with a chisel as you come to them so the cable will fit flush.

4. Run the wire in this manner around the doorway down to the newly located switch cut in the wall beside the door, or down and on to the next receptacle to be installed.

5. Replace the molding or casing around the door.

Running Cable Around a Doorway

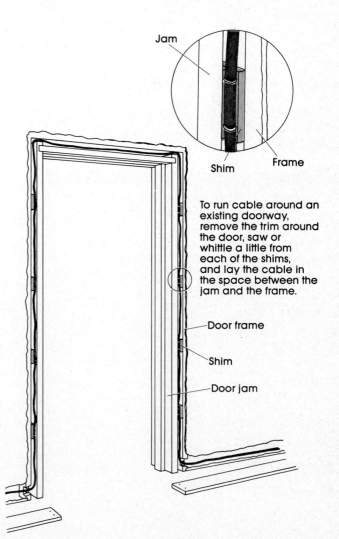

Jam

Shim

Frame

To run cable around an existing doorway, remove the trim around the door, saw or whittle a little from each of the shims, and lay the cable in the space between the jam and the frame.

Door frame

Shim

Door jam

Connecting Back-to-Back Boxes

Installing a new receptacle in a wall where there is already a receptacle located on the other side of the wall is one of the easier tasks in wiring. Because the wall is normally only as thick as the wall stud, the boxes will be mounted on two adjoining studs. There would not be room on the same stud.

Here are the steps involved:

1. After making sure the power is off, remove the wall plate and the receptacle. Leave the wires intact. Slip a coat hanger between the box and the wall to determine which side of the box the stud is on.

2. Measure from the edge of that stud to the nearest doorway. Then repeat that measurement on the other side of the wall to locate the same stud. The next stud normally will be 16 inches away from that one. In some modern houses, it may be 24 inches away but you can quickly determine this by rapping on the wall and listening for a solid sound.

3. Cut the opening for the new box beside the stud at the same height as the existing one.

4. With a screwdriver, remove the knockout in the back of the existing box and push the new wire through that and over to the new hole.

5. Install the new box. Clamp the cable in both boxes. Connect the cable to both receptacles.

Back-to-Back Boxes

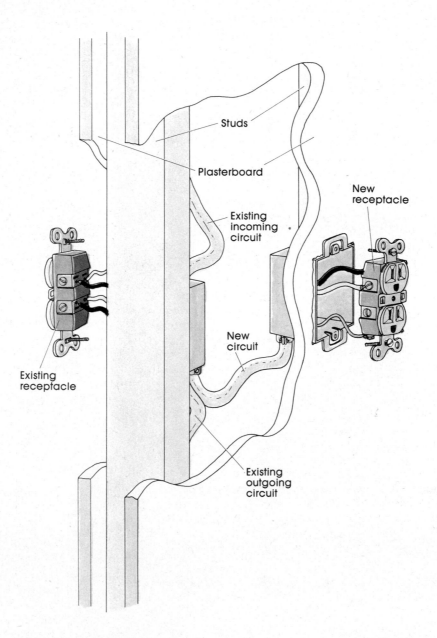

Studs

Plasterboard

New receptacle

Existing incoming circuit

New circuit

Existing receptacle

Existing outgoing circuit

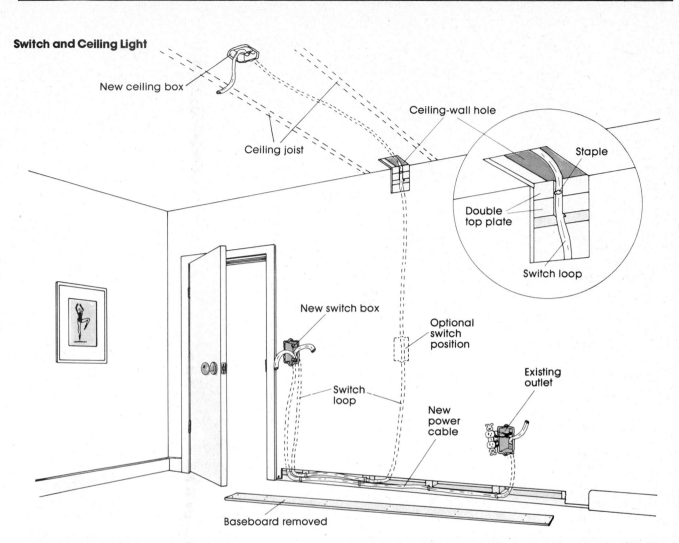

New ceiling box

Ceiling joist

Ceiling-wall hole

Staple

Double top plate

Switch loop

New switch box

Optional switch position

Existing outlet

Switch loop

New power cable

Baseboard removed

Adding a Switch and a Ceiling Light

This project is designed to add a new switch that controls a new ceiling light, all hooked into an existing receptacle in the room. For this example, it is to be done the hard way—with no access above the ceiling or under the house.

The first thing you must determine is which way the ceiling joists run. Do this by rapping on the ceiling and listening for the solid sound indicating the joist. Once you locate it, continue rapping along one direction that has a continuous solid sound, indicating the joist runs across the ceiling in that direction. When you fish the new cable above the ceiling, it must run parallel with the joists. It would be difficult to fish it across the tops of several joists.

If you locate the new switch directly in line with the joists that run to the light, you can then run the switch loop directly up from the switch. In order to do this, you may need to locate the switch in the middle. But it isn't that difficult to put the switch where you want it. Here is the step-by-step process.

1. Decide where you are going to put the light and cut a hole for the ceiling box (see page 69). Directly in line with the ceiling hole, where the ceiling meets the wall, cut a 4-inch-wide hole, as illustrated. This hole will be centered between the same two ceiling joists as the ceiling box.

2. Decide where you want the switch, usually next to a door, and cut a hole for the switch box.

3. Run a cable (as shown on pages 69–71) from the nearest receptacle to directly below the hole for the switch. This will be your power cable. Feed a fish tape from the switch hole down to the cable, attach the cable to the tape, and pull the cable up and out the switch hole. Let about 12 inches of the cable protrude from the hole.

4. Run a second cable, your switch loop, from the switch hole to a spot directly below the ceiling-wall hole that is in line with the light hole. Fish the switch-loop cable up to the ceiling-wall hole and then over to the ceiling hole. With a chisel, notch the double top plates in the wall so the cable will be recessed as it passes there into the ceiling. Staple the cable in the notch. Again, let about 12 inches of cable protrude from the hole. Your project should now look as shown in the drawing.

5. Connect the wires to the existing receptacle and to the switch and clamp the cable in the ceiling box.

6. In plasterboard walls and ceilings, patch the holes as described on page 74. In lath and plaster, fill the holes with plaster patching compound to about 1/8 inch below the wall surface. Let dry, and then smooth on a coat of spackling compound. Let dry, sand smooth, and paint.

OUTDOOR WIRING

Once you learn to tap a power source
for outside and extend a cable
underground, you can easily install
outdoor fixtures and low-voltage lights—
for convenience, safety, beauty, and
many practical uses.

Outdoor wiring can mean a change in your living style. It can turn a yard, garden, or patio that goes unused after nightfall into an outdoor living room or recreation area. It takes just a few strategically placed lights to transform an otherwise dark and unused yard. Put one in a tree, or place a series of low-voltage lights along a border garden or on posts around a deck to give the yard a whole new mood after dark. The added lights, not all of which have to be on at the same time, also help reduce chances of burglary.

Outdoor wiring is basically the same as indoor wiring. You use the same types of switches and outlets. The wiring connections also are done in the same manner. Two aspects that are different in outdoor wiring are: you must use waterproof boxes and the wiring must be resistant to water.

Planning Ahead

Before starting an outdoor wiring project, you can save yourself a great deal of unnecessary work by making a detailed plan.

First, decide how to take the electricity from inside to outside the house. You can tap existing circuits in the attic, the basement, or the house, all of which will be detailed later in the chapter.

Second, make a careful plan of the new lighting circuit route and what lighting effects you want developed. The best way to start is by making a scale drawing of your yard and house. Measure both the yard and house with a 100-foot steel tape and then transfer the measurements to ¼-inch graph paper. Each ¼-inch square will represent one foot. Don't forget to include all the walkways, trees, shrubs, border gardens, fences, and decks.

Make the scale drawing in ink and then, using a pencil, begin playing with ideas. Note the locations where you will connect the outdoor circuit to the house. You may see that you will need two or more such circuits, certainly one in front and one in back.

With a light pencil line, sketch where you want this circuit to be buried in order to serve all the required area with the least amount of wire. You may want light on steps and walkways for safety, perhaps a floodlight under the eaves as a burglary deterrent, and decorative lights along the deck or in trees and flower gardens. Once your plans are complete, you are ready to go to work.

Make a Detailed Plan

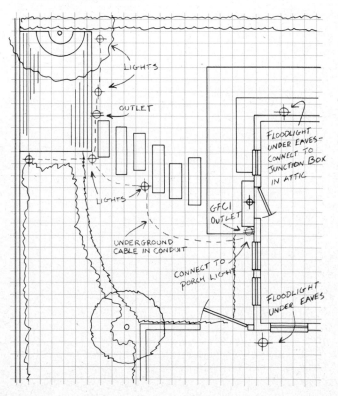

◀ **A well-planned outdoor lighting scheme can add interest to your landscape and make it safer.**

OUTDOOR WIRING DEVICES

Although wiring outdoors is the same in principle as wiring indoors, you need specialized weatherproof devices. The switches and receptacles used outdoors are the same as those used inside the house but the boxes are specially designed to be waterproof. Gaskets around the edges prevent water from leaking in, and spring-operated flaps, complete with gaskets, snap over receptacles and switch handles.

The lighting devices used outside must be specially designed to be leakproof. Even the bulbs used outdoors must have special glass that will not shatter from a sudden temperature change when hit by sleet or rain.

The cable used in outdoor wiring is also different. NM cable, commonly used inside the house, has fiber wrapped around the individual wires, which acts like a wick to draw moisture. Outside wiring must be weatherproof, such as NMC cable, TW wiring, or the completely waterproof UF (underground feeder) cable (see page 17).

Outdoor wiring involves the use of conduit, or pipe, to protect the wires. Although some local codes allow UF cable to be buried directly in the ground, it is generally best to encase it in conduit for maximum protection. NMC cable and Type TW wire must be run through conduit to be safe outdoors. Since Type TW is a wire, not a cable (two or more wires in the same sheathing), you must buy individual wires for the hot, neutral, and grounding wires and hook them up like cable.

Exterior Fixtures

Five basic fixtures, all highly weather-resistant, are used in outdoor wiring. They include the outdoor receptacle box, outdoor switch box, box extender, LB connector, and insulated bushing. The boxes are made of heavy aluminum or galvanized steel to resist rust. All openings, such as around the cover plates, are sealed against moisture by gaskets. Even the screw holes are countersunk and have gaskets to keep out moisture. Small spring-loaded doors with gaskets snap closed over the switches and receptacles. Instead of knockouts, all openings are covered by threaded plugs that can be removed for wires or conduit.

Outdoor receptacle box. This metal box consists of three main parts: the box, a cover-plate gasket, and cover plate. The cover plate has two spring-loaded doors with gaskets that close over the receptacle openings when not in use. Ordinary receptacles may be used in this box. The box has threaded plugs to seal openings when not used for wires or conduit. Although highly water-resistant, it should not be used where it might be submerged during sporadic flooding.

Outdoor switch box. This switch box is exactly the same as the receptacle box, complete with water-tight plugs that close unless used for wiring or conduit. Ordinary switches can be used in this box if you use the gasket and cover plate that has the spring-loaded door to protect the switch opening. However, a preassembled switch and a cover plate that screws on to the box are also available. A small handle outside the cover plate controls the switch inside by just moving it up or down.

Extender box. It is quite likely that your house, even if fairly old, has at least one receptacle on the exterior wall. These receptacles, however, are generally built flush with the exterior wall. To bring them out from the wall so you can easily tap them for your outdoor wiring, use an extender box, which is simply a switch box without a back that screws on to the existing box.

LB connector. The LB (L-shaped with a removable back) connector is used to run wires from inside the house through a hole in the wall and then down into the circuit. The holes at each end are threaded to accept fittings. A *nipple*, or short length of threaded pipe, is run through the hole in the wall and screws into the LB connector. The conduit pipe screws into the other end. A gasket fits under the cover plate to keep out moisture; the cover plate allows access to the wires.

Insulated bushing. When cable is run into a steel nipple or conduit, the rough edges may wear and cut the plastic sheathing around the cable. To prevent that, an insulated bushing, with a smooth rubber or plastic end piece, is screwed on the pipe at that point. The cable runs through a hole in the bushing.

Outdoor Wiring Fixtures

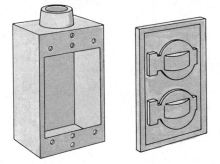

Outdoor receptacle box and cover

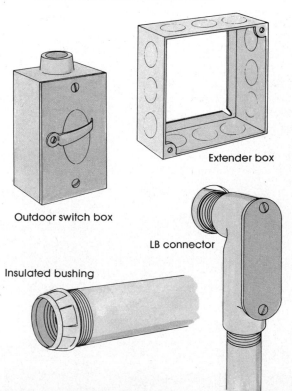

Extender box

Outdoor switch box

LB connector

Insulated bushing

Conduit

Conduit is pipe that contains and protects cable that is run outdoors, on the side of the house, or underground. Three types of conduit are commonly used, each requiring different techniques for joining it together. These are: plastic conduit (PVC), thinwall conduit (EMT), and rigid conduit, either steel or aluminum. Conduit differs from water pipe in that the interior is very smooth so that cable can be pulled easily and without damage to the insulation. In most cases, ½-inch-diameter conduit is sufficient. However, the size of conduit is determined by the number and size of the wires it contains.

Plastic conduit (PVC). Of the three types of conduit available, PVC (polyvinylchloride) pipe is the easiest to work with. It is lightweight, completely waterproof, and will not corrode. PVC commonly comes in 20-foot sections. On some types, one end of the pipe is belled and the next piece slips into that end; on others, individual connectors must be used to join each piece. PVC can be easily cut with a hacksaw. Use the edge of a knife blade to scrape away any burrs around the end of a freshly cut piece before joining it to another. This plastic pipe is glued in place with a special PVC glue that sets in seconds. Lightly sand the surfaces of both pieces to be glued, swab glue on both surfaces, and then push them together. Give a quarter turn to spread the glue and you're done. If one of the pieces to be glued must be aligned a certain way, such as an ell fitting at a corner, make sure you align it exactly. In ten seconds you won't be able to move it. When making right-angle turns with PVC, use two 45-degree ells in order to make a curving turn. It is easier to pull wire through that kind of corner than a direct 90-degree ell. Threaded couplings to fit all boxes are available with PVC pipe.

Conduit

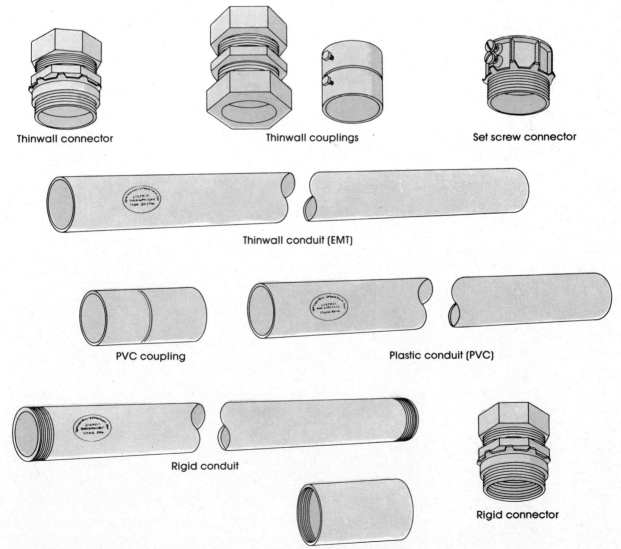

Thinwall connector

Thinwall couplings

Set screw connector

Thinwall conduit (EMT)

PVC coupling

Plastic conduit (PVC)

Rigid conduit

Rigid connector

Rigid coupling

Thinwall conduit (EMT). Thinwall conduit, or electrical metallic tubing, is made either of galvanized steel or aluminum. Some EMT is threaded on the end for couplings. On another type, the joints are held together with special couplings that hold the conduit in place by pressure. Be sure to ask for outdoor couplings, which are watertight. Thinwall is cut with a hacksaw, preferably with a blade having 32 teeth per inch for the smoothest cut. After the cut, the rough edges must be smoothed with a round file or a special pipe reamer.

Rigid conduit. Used for standard outdoor wiring underground, rigid conduit comes in 10-foot sections. The ends are threaded and joined with couplings. If you know the exact dimensions of your underground layout, the pipe can be cut and threaded to your specifications where you buy it. Otherwise, you must rent tools and do it yourself. Rigid pipe can be cut with a hacksaw, but if you have much to cut, it is worth it to rent a portable band saw with a metal cutting blade. Never use a pipe cutter, as used in cutting water pipe. It leaves the edges so sharp they may cut the insulation.

Bending Conduit

Thinwall and rigid conduit are normally assembled before the cable is fished through. However, with plastic conduit, all the pieces can be laid out and the cable can be run through each piece before it is glued together.

But since wires must be fished through after thinwall or rigid conduit is assembled, there can be no sharp bends that make it impossible to pull the cable around. The conduit must be bent in a gentle curve and at the same time not reduce the interior diameter. For this job you should rent a conduit bender. Hook one end over the pipe and then, holding the bender in place with your foot, pull the bender back until the pipe bends to the required angle.

Some pipe benders have spirit levels built into them so you can bend the conduit to precise 45- or 90-degree angles. You can still do quite accurate work without the spirit levels by just eyeballing the angle of bend.

Burying Conduit

How deep conduit should be buried may differ according to local codes, but use common sense. If the conduit is going to be in an area where the ground may be worked with a shovel or rototiller, put it at least 12 inches down to be out of harm's way. It's also a good idea to cover it with lengths of 2×4 redwood or treated wood that will not rot underground.

In many areas, codes permit rigid conduit to be buried only 6 inches deep because the steel pipe provides strong protection. PVC and UF cable must be buried deeper, down to 18 inches in many areas. Check your local codes on this matter.

Using a Conduit Bender

1. Insert the conduit in the bender so the hook is where you want the bend to start.

2. Hold the bender and conduit in place with your feet and pull the handle back until the conduit is bent to the required angle.

TAKING ELECTRICITY OUTSIDE

When your outdoor wiring plans are complete, the first step is to take a source of power from inside the house to the outside. The process is quite similar to extending a circuit inside the house, as discussed on pages 66–68. Basically it means tapping into a receptacle or an exposed circuit and installing a junction box in the house, then extending the new cable through the wall to the house exterior. Here are several ways to install a power source for outdoor lighting.

With an Extender Box

If your house already has a flush-mounted receptacle on the outside of the house, it is a fairly simple process to connect your new outdoor cable to that. The job is done with an extender box. Shut off power on that circuit and remove the faceplate. Pull the receptacle out of the box but leave the wires intact. Screw the extender box onto the existing box. Unscrew the plug in the bottom of the extender box. Bring the new cable up through the conduit and wire it into the receptacle. Screw the conduit into the threaded hole in the extender box and install the cover plate over the extender box.

With Back-to-Back Boxes

This system is similar to installing a new box on the opposite side of a wall, as described on page 63. Shut off power to the receptacle you are going to tap inside the house. Remove the wall plate and the receptacle from the box, leaving the wires intact. Punch out the rear knockout in the box. If the box is plastic, drill a ½-inch hole in the back. Drill a ½-inch hole through the wall to the house exterior. Remove the rear knockout on the cast metal box for the outside wall and then screw it to the house with the knockout hole centered over the hole in the wall. Push a length of cable from inside to outside, leaving about 8 inches hanging out each side. Connect the wires to the receptacle inside and to the receptacle or switch outside. Replace the cover plates on both boxes and then put a bead of caulking compound around the box where it meets the house.

Installing a Back-to-Back Box Outside

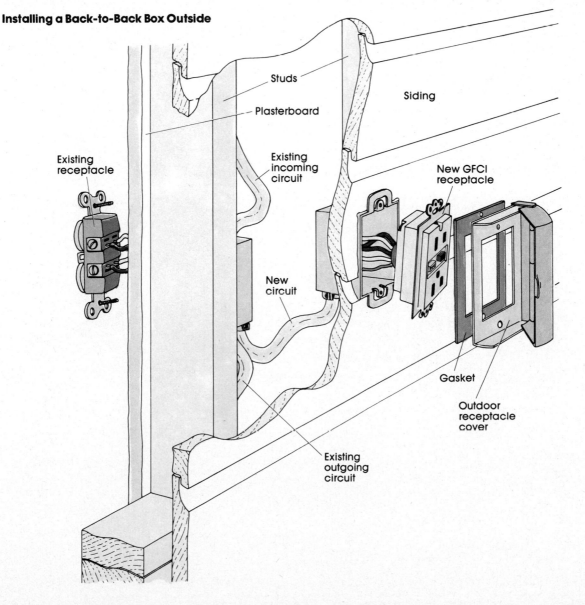

Studs

Plasterboard

Siding

Existing receptacle

Existing incoming circuit

New circuit

New GFCI receptacle

Gasket

Outdoor receptacle cover

Existing outgoing circuit

From the Attic or Basement

Wires may often be readily accessible in the attic or basement, particularly in older houses. Locate an available circuit near the wall you want to go through. Then make sure power is off on that circuit. If you have doubts, shut off all power in the house. Cut the wires and install a junction box at that point, as described on page 67. Connect the new circuit to the existing one (black to black, white to white) with wire nuts.

Drill a ⅞-inch hole through the wall with a spade bit. Buy a ½-inch-diameter nipple (short length of pipe threaded at both ends) long enough to extend through the wall. Fit the interior end with an insulated bushing to protect the wires. Screw the other end of the nipple to an LB connector on the exterior wall. Join the other end of the LB connector to the conduit that runs to the switch or receptacle box conveniently located on the side of the house.

Tapping a Porch Light

Another way to establish an outdoor circuit is to tap into the front or back porch light on your house. This requires a box extender, some LB fittings, and some EMT conduit.

The first step is to shut off power to the porch light circuit and remove the light from its mountings. Remove the wires from the light.

Next, screw a short nipple and an LB fitting, or corner elbow, to the threaded hole in the bottom of the box extender. Attach the box extender to the existing box on the house wall and then run cable through the nipple and corner elbow.

Taking a Circuit Outside

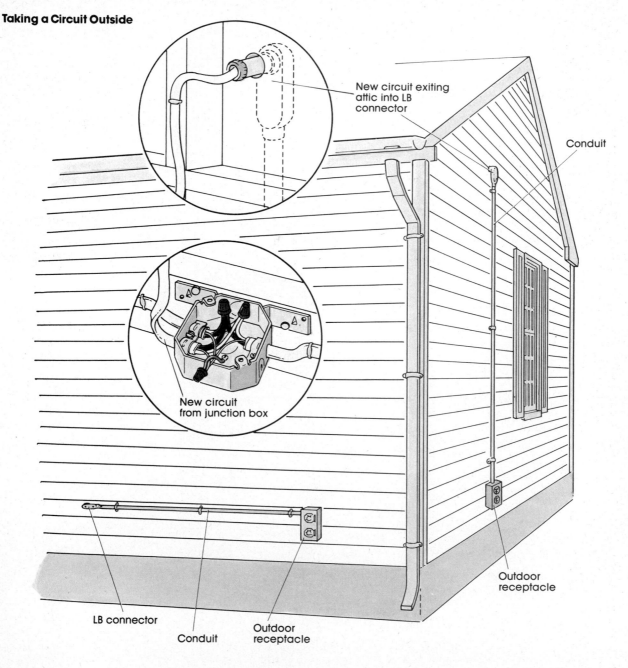

New circuit exiting attic into LB connector

Conduit

New circuit from junction box

Outdoor receptacle

LB connector

Conduit

Outdoor receptacle

Installing an Extender Box on the Porch Light

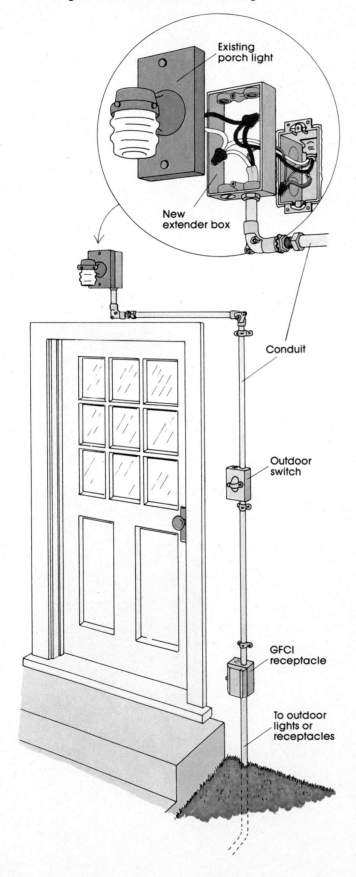

Existing porch light

New extender box

Conduit

Outdoor switch

GFCI receptacle

To outdoor lights or receptacles

Cut a length of EMT conduit long enough to reach just beyond the width of the porch or steps and install another corner elbow on the end of that. Run the cable through his segment, connect the conduit and fasten it to the house with conduit strap.

If you want a switch, run another length of conduit down to the height where you want to locate the switch. Pull the cable through the conduit to that point and install an exterior switch box and switch there.

If the porch light circuit is not already protected by a GFCI at the service panel (which is not likely), the Code requires you to install one on this outdoor circuit. Since the CFCI/receptacle combination will protect all other outlets and fixtures along that circuit, it should be the first one installed. A good place to put is is right beside or below the switch that controls the outdoor lights. The receptacle part of the GFCI located here beside the porch, deck, or patio will prove handy for plugging in additional lights or electric cooking devices. Follow the instructions for the particular model of GFCI you purchased.

From that point, run conduit down the side of the house and into the prepared ditch. Fasten it to the house with conduit straps. At the ditch, the conduit can be bent with a conduit bender (before fastening to the house) or you can use another LB fitting to make the turn into the ditch.

Once the wire has been pulled through the conduit and connected to all the other outdoor receptacles and fixtures, you are ready to connect the cable to the porch light.

If there are not enough screw terminals on the light for the six wires (three incoming and three outgoing hot, neutral, and ground wires), use wire nuts and pigtails to make the connection as illustrated.

Safety Reminders

There are a few points to keep in mind when working with outdoor wiring:

■ Make sure that power is off before attempting to work on a circuit. Because damp ground makes such an efficient conductor, be doubly cautious outside. Shut off the circuit even for such a simple task as changing an outdoor light bulb.

■ All outdoor circuits must be protected by a ground fault circuit interrupter (GFCI). Damp grass or ground makes just about as good a conductor as the copper grounding wire in the circuit. Thus, if there is a short circuit, electricity will pass through your body to the damp ground just as readily as it will through the grounding wire, and the results could be fatal. If the GFCI senses any change in the amount of electricity flowing through the wires, it shuts off the circuit in about a fortieth of a second, fast enough to prevent any serious harm to you.

■ If you have plans to wire a pool or hot tub, leave it to a professional. The wiring in such a project can be fairly intricate and you don't want any mistakes in this area.

Extending Cable Underground

Once you have extended a source of power to the house exterior, you are ready to install the underground wiring system. This basically involves digging the ditch and installing any switches, receptacles, or lights along the route.

Starting from directly under the exterior power source, lay out the trench line with stakes and string. Mark all locations for additional switch boxes or lights. Mark the

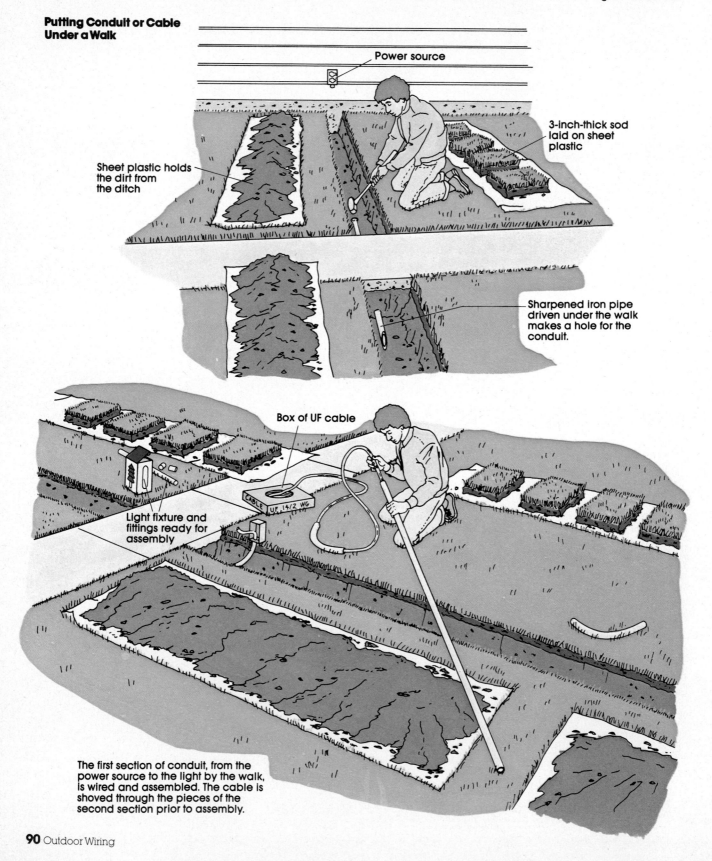

Putting Conduit or Cable Under a Walk

Power source

Sheet plastic holds the dirt from the ditch

3-inch-thick sod laid on sheet plastic

Sharpened iron pipe driven under the walk makes a hole for the conduit.

Box of UF cable

Light fixture and fittings ready for assembly

The first section of conduit, from the power source to the light by the walk, is wired and assembled. The cable is shoved through the pieces of the second section prior to assembly.

layout on the grass with chalk or flour and then remove the strings. The ditch need only be as wide as your shovel, which will also be wide enough for you to work in comfortably.

Lay out 2-foot-wide sheets of plastic on each side of the ditch. On one sheet put the sod, which should be cut about 3 inches deep to minimize root damage. On the other sheet, pile the dirt from the trench. From the plastic, the dirt can be readily put back in the trench without leaving a trail of dirt in the grass.

When you come to a sidewalk, dig the trench to one side and then continue on the other side. Dig out underneath the sidewalk as much as you can with the shovel. Then take a length of ½-inch steel pipe about half again as long as the sidewalk is wide and cut one end at an angle to form a point. Hammer that end flat and then use a sledgehammer or five-pound maul to drive the pipe through the intervening dirt.

When the ditch is done, lay out all the conduit beside the trench to see that it matches. Connect a long section with no bends in it and push the cable through. Push the cable through any curved conduit, then connect the two pieces of conduit. Continue in this fashion through the conduit layout. Hook up the cable to switchboxes or lights as you come to them.

Even if you use UF cable instead of underground conduit, remember that you must have conduit running from the box on the side of the house down into the ground to protect the exposed wires.

Installing Outdoor Fixtures

After the ditch is completed and the conduit laid out beside it, you are ready to begin installing the outdoor fixtures, such as lights, switches, and receptacles. From your plans, you should know where they will be most advantageously placed.

One of the easiest and least expensive means to securely anchor conduit where it emerges from the earth for a fixture or switch box is with cinder building blocks. When the wires have been pulled through the conduit to that point and the conduit all connected, slip a cinder block over the end of the conduit at that point and set it in the ditch. The curve in the conduit there can be made with a conduit bender or by using LB fittings.

Use a carpenter's level to make sure the conduit is vertical, then fill the hole in the cinder block around the conduit with ready-mix concrete. This is available at most hardware or lumber stores; just add water, mix, and pour into the cinder block.

For a middle-of-the-run connection, use an outdoor junction box with three knockouts. Remove two knockouts in the bottom for the arriving and departing conduit and use one in the top for the fixture stand, as illustrated. In the junction box, use wire nuts to join the wires—black to black, white to white, and ground to ground.

For the end-of-the-run connection, use a similar three-hole box if you want two receptacles located there, as illustrated. If you need only one fixture or single outlet at the end, use a single switch box.

Outdoor Fixtures

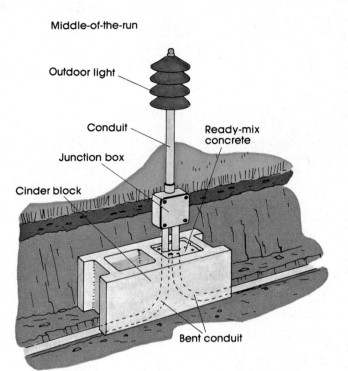

Middle-of-the-run

Outdoor light

Conduit

Ready-mix concrete

Junction box

Cinder block

Bent conduit

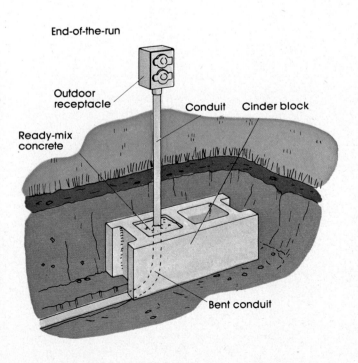

End-of-the-run

Outdoor receptacle

Conduit   Cinder block

Ready-mix concrete

Bent conduit

LOW-VOLTAGE LIGHTS

Low-voltage lights are increasing in popularity for outdoor lighting. Since they use only 12 volts instead of standard 120 volts, you get decorative lighting for less energy and cost. They are also very easy to install.

Low-voltage lighting systems can be bought in kit form, which primarily include a certain number of lights and a transformer. The transformer reduces the incoming 120-volt power down to 12 volts. Transformers differ in size according to how many lights you are going to use.

The transformer comes in a water-tight box that is mounted on the house near the exterior power source. To hook it up, shut off power on that exterior circuit and connect the two boxes with a short length of conduit. Run a length of No. 12 cable from the power source to the transformer and hook it up according to the instructions on that particular model.

From the transformer, run a length of UF cable along the voltage lighting route. This wire size is specified in the instructions and depends on how many lights you are using. This cable can be buried only a few inches in the ground because, if cut, the low voltage would not harm anyone. However, place it where it is not likely to be severed by a shovel. The cable hooks into the lights by clips in some cases or by standard wiring methods in others.

Low-Voltage Lighting Installation

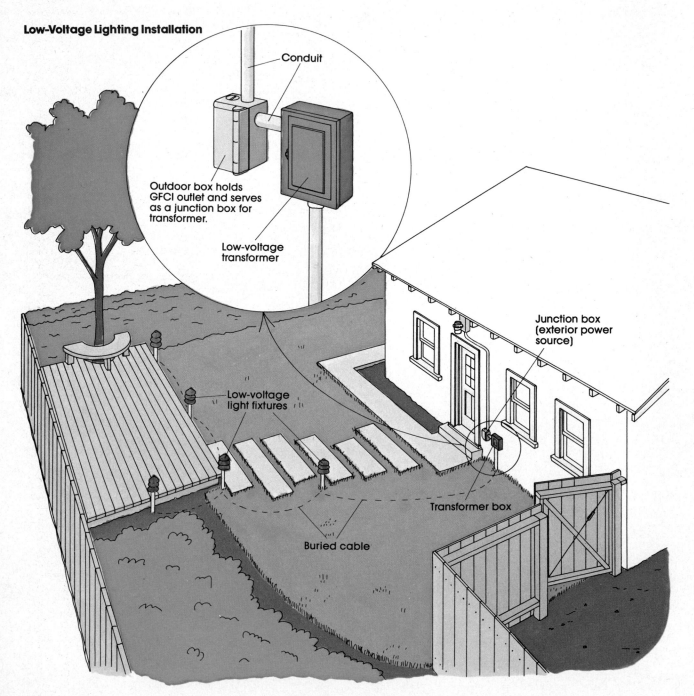

Conduit

Outdoor box holds GFCI outlet and serves as a junction box for transformer.

Low-voltage transformer

Junction box (exterior power source)

Low-voltage light fixtures

Transformer box

Buried cable

GLOSSARY

Alternating current (AC). An electric current in which voltage flows in one direction one instant and in the other direction the next. The direction of current flow reverses regularly.

Ampacity. How much current, expressed in amperes, a conductor can carry.

Ampere. A measure of the amount of electrical current flow.

Approved. This means that minimum standards established by an authority have been met.

Arc, electric. A visible, sustained discharge of electricity that bridges a gap in a circuit or between electrodes.

Armored cable. A flexible, metallic-sheathed cable used for indoor wiring. Commonly called BX.

Ballast. A magnetic coil that adjusts current through a fluorescent tube, providing the current surge to start the lamp.

Bonding. The permanent adhesion of metallic parts, forming an electrically conductive path. *See also* Ground.

Box. A wiring device that is used to contain wire terminations where they connect to other wires, switches, or outlets.

Branch circuit. A circuit that supplies a number of outlets for lights or appliances.

Burrs (conduit). The rough metal edges caused by cutting conduit pipe.

Bus bar. A heavy solid conductor at the main power source to which branch circuits are connected.

BX. *See* Armored cable.

Cable. A conductor, consisting of two or more wires that are grouped together in an overall covering.

Candlepower. The intensity of a light beam.

Circuit. The path of electric current, leading from a source (generator or battery) through components (for example, electric lights), and back to the source.

Circuit breaker. An electromagnetic or thermal device that opens the circuit when the current in the circuit exceeds a predetermined amount; can be reset.

Code, National Electric (NEC). A set of rules sponsored by the National Fire Protection Association, under the auspices of the American National Standards Institute, to protect persons, buildings, and contents from dangers due to the use of electricity.

Color coding. The identification of conductors by color.

Common ground connection. The place where two or more continuous grounded wires terminate.

Conductor. A low-resistance material, such as copper wire, which electricity flows through easily.

Conductor, bare. A conductor that has no covering or insulation.

Conduit. A metal or fiber pipe or tube that is used to enclose electrical conductors.

Connector, solderless. A device that uses mechanical pressure, rather than solder, to establish a connection between two or more conductors.

Continuity. An uninterrupted electrical path.

Current. The movement or flow of electrons; the time rate of electron flow, measured in amperes.

Cycle. A complete positive and negative alternation of a current or voltage.

Device. A unit or component that carries but doesn't use current, such as a junction box or switch.

Direct current (DC). An electric current that flows only in one direction.

Electrical charge. The electrical energy of a body or particle. The electron has an inherent negative charge; the proton has an inherent positive charge.

Electric current. The flow of electrons through a conductor.

Electron. The negatively charged particle of an atom. The flow of electrons in a conductor is what constitutes electric current.

Exposed. Wiring that is designed to be easily accessible.

Feeder. The circuit conductors between the service equipment and the branch circuit overcurrent device.

Fish tape. Flat, steel spring wire with hooked ends; used to pull wires through conduits or walls.

Fitting. An accessory (such as a bushing or locknut) used on wiring systems to perform a mechanical, not electrical, function.

Flexible cable (flexible cord). A conductor that is made of several strands of small-diameter wire.

Footcandle. A measurement of light; the amount of illumination that occurs when 1 lumen falls on 1 square foot of surface.

Fuse. A safety device that is inserted in series with a conduit, containing metal that will melt or break when the current exceeds a specific value for a specific time period.

Ground. A connection between an electrical circuit and the earth or a body serving in place of the earth.

Grounded. Connected to the earth or some body serving in place of the earth. The grounded wire is always white.

Grounding conductor. The wire (green) in a cable that carries no current; used as a safety measure.

Hanger. A metal or insulated strap used at intervals to support electrical cable between points of connection.

Hot wires. The wires (any color but white and green) of a house circuit that are not connected to ground and are carrying power.

Impedance. The total opposition to alternating current created by an electrical circuit.

Insulation. Materials that do not carry current; used on the outside of wires and in the construction of electrical devices.

Insulator. A nonconductor that is used to support a conductor that carries current.

Junction box. A box in which several conductors (wires) are joined together.

Knockout. A circular metal die-cut impression in outlet and switch boxes that has not been completely severed and that may be removed to accommodate wiring.

Lighting outlet. An outlet meant to allow the direct connection of a lampholder, lighting fixture, or pendant cord ending in a lampholder.

Live wire. A wire that carries current.

Meter, electric. A device that measures how much electricity is used.

National Electric Code. *See* Code, National Electric.

Neutral wire. The wire in a cable that is kept at zero voltage. All current that flows through the hot wire also must flow through the neutral wire.

Nonconductors. Materials that electric current does not flow through (such as glass, porcelain, rubber).

Ohm. The amount of electrical resistance in a circuit or electrical device.

Open circuit. An electrical circuit with a physical break in the path (caused by opening a switch, disconnecting a wire, burning out a fuse, and so forth), through which no current can flow.

Outlet. A metallic or fiber box in which electrical wiring is connected to electrical components.

Overcurrent protection device. A fuse or circuit breaker that is used to prevent the excessive flow of current.

Overload. Current demand exceeding that for which the circuit or equipment was designed.

Parallel circuit (parallel connection). A circuit that provides more than one path for electrical current to flow.

Plug, attachment. A device that is inserted in a receptacle to establish a connection between conductors of attached flexible cord and conductors that are permanently connected to the receptacle.

Polarized plug. A plug whose blades are designed to enter a receptacle in only one orientation.

Polarizing. The use of color to identify wires throughout a system to make sure that hot wires will be connected only to hot wires, and that neutral wires will run back to the ground terminals in continuous circuits.

Power. The rate at which work is being done. The unit of electrical power measurement is the watt.

Raceway. A channel that supports electrical conductors (wires or cables).

Receptacle. A contact device that is installed at the outlet to supply current to a single extension cord plug.

Receptacle outlet. An outlet where one or more receptacles are installed.

Relay. An electrical component that opens or closes contacts via remote-control current application to an electromechanical element, thus opening and closing circuits.

Resistance. The quality of an electric circuit, measured in ohms, that resists the flow of current because of the conductor's atomic nature. Good conductors (such as copper, silver, aluminum) offer little resistance; poor conductors (such as glass, wood, paper) offer much resistance.

Romex. The commercial name for nonmetallic-sheathed electrical cable used for indoor wiring.

Screw terminal. A means for connecting wiring to devices, using a threaded screw.

Series circuit. An electrical circuit allowing only a single path for current to flow.

Service conductors. Conductors that extend from street main or transformer to the service equipment of the building being supplied with electrical service.

Service equipment. Equipment that is located near the entrance of supply conductors that provides main control and enables cutoff (fuses or circuit breaker) for the supply of current to the building.

Service panel. The main panel through which electricity is brought into a building and distributed to the branch circuits.

Short circuit. An improper connection between hot wires, or between a hot wire and a neutral wire.

Solderless connector (wire nut). A mechanical device (typically, plastic-insulated), that can be fastened over the exposed and joined ends of several wires to make a firm connection between those wires.

Splice. A connection that is made by connecting two or more wires.

Split receptacle. A dual receptacle in which each of the two receptacles is connected to a different branch circuit rather than to a common circuit.

Stranded wire (stranded cable). A quantity of small conductor wires that are twisted together to form a single conductor.

Switch. A device that is used to connect and disconnect the flow of current or to divert current from one circuit to another; used only in hot wires, never in ground wires.

Symbols, electrical. Line, letters, and so forth used on house plans to show where wiring circuits, switches, outlets, and so on are to be installed.

System, electrical. An electrical installation that is complete and will serve its intended purpose.

Three-way switch. A type of switch. Two of these switches are needed for controlling a light from two different locations.

Underwriters' knot. A knot that is used to tie two insulated conductors at the terminals inside an electric plug; used to relieve strain on the terminal connection.

Underwriters label (UL label). A label that is applied to manufactured devices that have been tested for safety by Underwriters Laboratories and approved for placement on the market. These labs are supported by insurance companies, manufacturers, and other parties interested in electrical safety.

Volt. A unit that is used to measure electrical pressure (comparable to pounds of pressure in a water system).

Voltage. The electromotive force or potential difference between two points of a circuit, measured in volts, that causes electric current to flow. One volt creates a current of 1 ampere through a resistance of 1 ohm.

Voltage drop. A loss of electrical current, caused by overloading wires or by using excessive lengths of undersize wire. Often indicated by dimming of lights and slowing down of motors.

Volt-ampere. In an AC circuit, a unit of measurement of electrical power equal to the product of volts times amperes. In DC, 1 volt-ampere equals 1 watt, and in AC it is a unit of apparent power.

Watt. A unit of measurement of electric power. (Volts times amperes equals watts of electrical energy used.) One watt used for 1 hour is 1 watt hour; 1,000 watt hours equals 1 kilowatt hour (the unit by which electricity is metered and sold by utility companies).

Wire. An electrical conductor in the form of a slender rod.

Wire gauge. A standard numerical method of specifying a conductor's physical size. The American Wire Gauge (AWG) series is most common.

Wire nut. *See* Solderless connector.

Wiring diagram. A drawing, in symbolic form, showing conductors, devices, and connections.

INDEX